P9-DIW-941

ANNAPOLIS MIDDLE SCHOOL
MEDIA CENTER

Life in the Universe

JOSEPH A. ANGELO, JR.

Facts On File
An imprint of Infobase Publishing

To my wife, Joan, in celebration
of our 42nd wedding anniversary

✧

LIFE IN THE UNIVERSE

Copyright © 2007 by Joseph A. Angelo, Jr.

All rights reserved. No part of this book may be reproduced or utilized in any form or by any means, electronic or mechanical, including photocopying, recording, or by any information storage or retrieval systems, without permission in writing from the publisher. For information contact:

Facts On File, Inc.
An imprint of Infobase Publishing
132 West 31st Street
New York NY 10001

Library of Congress Cataloging-in-Publication Data
Angelo, Joseph A.
   Life in the Universe / Joseph A. Angelo, Jr.
      p. cm.—(Frontiers in space)
   Includes bibliographical references and index.
   ISBN-10: 0-8160-5776-1
   ISBN-13: 978-0-8160-5776-4
   1. Life on other planets. 2. Outer space—Exploration. 3. Exobiology. I. Title.
   QB54. A5235 2007
   576.8'39—dc22   2006034860

Facts On File books are available at special discounts when purchased in bulk quantities for businesses, associations, institutions, or sales promotions. Please call our Special Sales Department in New York at (212) 967-8800 or (800) 322-8755.

You can find Facts On File on the World Wide Web at
http://www.factsonfile.com

Text design by Erika K. Arroyo
Cover design by Salvatore Luongo

Printed in the United States of America

VB KT 10 9 8 7 6 5 4 3 2 1

This book is printed on acid-free paper.

Contents

Preface

It is difficult to say what is impossible, for the dream of yesterday is the hope of today and the reality of tomorrow.

—Robert Hutchings Goddard

Frontiers in Space is a comprehensive multivolume set that explores the scientific principles, technical applications, and impacts of space technology on modern society. Space technology is a multidisciplinary endeavor, which involves the launch vehicles that harness the principles of rocket propulsion and provide access to outer space, the spacecraft that operate in space or on a variety of interesting new worlds, and many different types of payloads (including human crews) that perform various functions and objectives in support of a wide variety of missions. This set presents the people, events, discoveries, collaborations, and important experiments that made the rocket the enabling technology of the space age. The set also describes how rocket propulsion systems support a variety of fascinating space exploration and application missions—missions that have changed and continue to change the trajectory of human civilization.

The story of space technology is interwoven with the history of astronomy and humankind's interest in flight and space travel. Many ancient peoples developed enduring myths about the curious lights in the night sky. The ancient Greek legend of Icarus and Daedalus, for example, portrays the age-old human desire to fly and to be free from the gravitational bonds of Earth. Since the dawn of civilization, early peoples, including the Babylonians, Mayans, Chinese, and Egyptians, have studied the sky and recorded the motions of the Sun, the Moon, the observable planets, and the so-called fixed stars. Transient celestial phenomena, such as a passing comet, a solar eclipse, or a supernova explosion, would often cause a great deal of social commotion—if not outright panic and fear—because these events were unpredictable, unexplainable, and appeared threatening.

It was the ancient Greeks and their geocentric (Earth-centered) cosmology that had the largest impact on early astronomy and the emergence of Western Civilization. Beginning in about the fourth century B.C.E., Greek philosophers, mathematicians, and astronomers articulated a geocentric model of the universe that placed Earth at its center with everything else revolving about it. This model of cosmology, polished and refined in about 150 C.E. by Ptolemy (the last of the great early Greek astronomers), shaped and molded Western thinking for hundreds of years until displaced in the 16th century by Nicholas Copernicus and a heliocentric (sun-centered) model of the solar system. In the early 17th century, Galileo Galilei and Johannes Kepler used astronomical observations to validate heliocentric cosmology and, in the process, laid the foundations of the Scientific Revolution. Later that century, the incomparable Sir Isaac Newton completed this revolution when he codified the fundamental principles that explained how objects moved in the "mechanical" universe in his great work *Principia Mathematica*.

The continued growth of science over the 18th and 19th centuries set the stage for the arrival of space technology in the middle of the 20th century. As discussed in this multivolume set, the advent of space technology dramatically altered the course of human history. On the one hand, modern military rockets with their nuclear warheads redefined the nature of strategic warfare. For the first time in history, the human race developed a weapon system with which it could actually commit suicide. On the other hand, modern rockets and space technology allowed scientists to send smart robot exploring machines to all the major planets in the solar system (including the dwarf planet Pluto), making those previously distant and unknown worlds almost as familiar as the surface of the Moon. Space technology also supported the greatest technical accomplishment of the human race, the Apollo Project lunar landing missions. Early in the 20th century, the Russian space travel visionary Konstantin E. Tsiolkovsky boldly predicted that humankind would not remain tied to Earth forever. When astronauts Neil Armstrong and Edwin (Buzz) Aldrin stepped on the Moon's surface on July 20, 1969, they left human footprints on another world. After millions of years of patient evolution, intelligent life was able to migrate from one world to another. Was this the first time such an event has happened in the history of the 14-billion-year-old universe? Or, as some exobiologists now suggest, perhaps the spread of intelligent life from one world to another is a rather common occurrence within the galaxy. At present, most scientists are simply not sure. But, space technology is now helping them search for life beyond Earth. Most exciting of all, space technology offers the universe as both a destination and a destiny to the human race.

Each volume within the Frontiers in Space set includes an index, a chronology of notable events, a glossary of significant terms and concepts, a helpful list of Internet resources, and an array of historical and current print sources for further research. Based upon the current principles and standards in teaching mathematics and science, the Frontiers in Space set is essential for young readers who require information on relevant topics in space technology, modern astronomy, and space exploration.

Acknowledgments

I wish to thank the public information specialists at the National Aeronautics and Space Administration (NASA), the National Oceanic and Atmospheric Administration (NOAA), the United States Air Force (USAF), the Department of Defense (DOD), the Department of Energy (DOE), the National Reconnaissance Office (NRO), the European Space Agency (ESA), and the Japanese Aerospace Exploration Agency (JAXA), who generously provided much of the technical materials used in the preparation of this set. Acknowledgment is made here for the efforts of Frank Darmstadt and other members of the editorial staff at Facts On File whose diligent attention to detail helped transform an interesting concept into a series of publishable works. The support of two other special people merits public recognition here. The first individual is my physician, Charles S. Stewart III, M.D., whose medical skills allowed me to complete the set successfully. The second individual is my wife, Joan, who has provided for the past 42 years the loving spiritual and emotional environment so essential in the successful completion of any undertaking in life, including the production of this set.

Introduction

....

L *ife in the Universe* provides a space age examination of the basic
question: Is life, especially intelligent life, unique to Earth? Human
interest in the origins of life and the possibility of life on other worlds
extends back deep into antiquity. Throughout history, each society's "creation myth" seemed to reflect that particular people's view of the extent of
the universe and their place within it. Today, as a result of space technology,
the scope of those early perceptions has expanded well beyond the reaches
of this solar system—to other stars of the Milky Way galaxy, to the vast
interstellar clouds that serve as stellar nurseries, and beyond to numerous
galaxies that populate the seemingly infinite expanse of outer space.

Just as the concept of biological evolution implies that all living organisms have arisen here on Earth by divergence from a common ancestry, so
too the concept of cosmic evolution implies that all matter in a solar system has a common origin—a primordial cloud of dust and gas with roots
that extend all the way back to the big bang event that occurred about 14
billion years ago. This book describes how scientists now use the overarching concept of cosmic evolution to postulate that life may be viewed as
the product of countless changes in the form of primordial stellar matter—changes brought about by the interactive processes of astrophysical,
cosmological, geological, and biological evolution.

Life in the Universe shows how the arrival of space technology, as well
as enormous space age improvements in ground-based astronomy, are
being used to hunt for life, existent or extinct, on other worlds in the solar
system. The Red Planet, Mars, and the intriguing Jovian moon, Europa, are
currently prime candidates for expanded investigation. This book explains
the principles of exobiology and how these principles are being used to
guide robot spacecraft in the search for life beyond Earth. Exobiology (also
called astrobiology) is the multidisciplinary field that involves the study
of extraterrestrial environments for living organisms, the recognition of
evidence of the possible existence of life-forms in these environments, and
the study of any nonterrestrial life-forms that may be encountered. One
key concern, if nonterrestrial life is found in microscopic form somewhere

in the solar system, is that of extraterrestrial contamination. *Life in the Universe* describes the international planetary quarantine procedures that have been developed and used by life-hunting scientists within the space exploration community.

Recent astronomical evidence suggests that planet formation is a natural part of stellar evolution. So scientists are now using a variety of Earth-based and space-based techniques in their continuing search for extrasolar planets, especially Earth-like planets that may have the capacity to support life. If life originates on "suitable" planets whenever it can (as many exobiologists currently suggest), then knowing how abundant such suitable planets are in the Milky Way galaxy would allow scientists to make more credible guesses about where to search for extraterrestrial intelligence and what the basic chances are of finding intelligent life beyond humans' own solar system. *Life in the Universe* includes some of the well-known speculative discussions about intelligent extraterrestrial life, including Kardashev civilizations, the Dyson sphere, the Fermi Paradox, and the Drake equation (see glossary).

This book also describes some space age efforts at interstellar communication, including the Arecibo radio telescope message, the special plaques placed on the *Pioneer 10* and *11* spacecraft, and the digital recordings carried by the *Voyager 1* and *2* spacecraft. No discussion of intelligent alien life is complete without some additional speculation concerning the consequences of contact and the societal impact of finding out that the human beings may not be the only intelligent inhabitants of the galaxy. By addressing these and many other intriguing questions that have puzzled people since ancient times, *Life in the Universe* prepares the reader for some of the most exciting consequences that space technology may yield during this century.

Life in the Universe also describes the historic events, scientific principles, and technical developments that allow sophisticated robot exploring machines to visit intriguing worlds in the solar system, hunting for signs of life—existent or extinct. The book's special collection of illustrations includes historic, contemporary, and future extraterrestrial life-searching robot spacecraft—allowing readers to appreciate the tremendous aerospace engineering progress that has occurred since the dawn of the space age and what lies ahead. A generous number of sidebars are strategically positioned throughout the book to provide expanded discussions of fundamental scientific concepts and speculative theories about alien life. There are also capsule biographies of scientists who pioneered various aspects of exobiology.

It is especially important to recognize that throughout this century and beyond, the space technology-based search for life beyond Earth could produce many exciting scientific discoveries that have enormous consequences for the human race. Such awareness should prove career inspiring to those students, now in high school and college, who will become the exobiologists, planetary scientists, or aerospace engineers of tomorrow. Why are such career choices important? If consciousness and life proves to be extremely rare, then future generations of human beings have a serious obligation to the entire (still "unconscious") universe to preserve carefully the precious biological heritage that has taken about four billion or so years to emerge and evolve here on planet Earth. If, on the other hand, life (including intelligent life) is found to be rather abundant throughout the galaxy, then future human generations might eagerly seek to learn of its existence and ultimately become part of a galactic family of conscious, intelligent beings. Sometime in the distant future, as a smart exploration robot travels to investigate a particularly interesting extrasolar planet around an alien sun, people here on Earth might finally be able to answer scientifically the age-old philosophical question: Are we alone in this vast universe?

Life in the Universe shows that the modern search for life beyond Earth did not occur without technical problems, political issues, or widely fluctuating financial commitments. Selected sidebars within the book address some of the most pressing contemporary issues that are associated with this search—including the long-standing space program debate about planetary contamination and the volatile political question: "Who speaks for Earth?"—if and when contact is ever made with an intelligent alien civilization. Some scientists and political leaders regard exobiology and the search for extraterrestrial intelligence (SETI) as rather meaningless scientific efforts "without *real* subjects." Other scientists accept exobiology and SETI as logical extensions of space age developments. *Life in the Universe* helps the reader make some informed choices concerning this search—the successful outcome of which would exert a tremendous influence on the trajectory of human civilization.

Life in the Universe has been carefully crafted to help any student or teacher who has an interest in the search for life beyond Earth to discover what efforts are taking place, how they work, and why they are potentially so important. The back matter contains a chronology, glossary, and an array of historical and current sources for further research. These should prove especially helpful for readers who need additional information on specific terms, topics, and events concerning the possibility of life beyond Earth and how space technology is influencing the search for it.

Alien Life: From Science Fiction to Mars Rocks

From the dawn of history, astronomical observations have played a major role in the evolution of human cultures. The modern human's distant ancestors viewed the sky and tried to interpret the mysterious objects that they saw.

For example, about 60,000 years ago, Neanderthal hunters must have gazed up into the night sky and wondered what the relatively large, bloodred-colored light was. (At that time, Mars was making one of its closest orbital approaches to Earth, so the ancient night sky must have been especially interesting.) The Neanderthals lived in the northern and

This artist's rendering shows Earth approximately 60,000 years ago. The elder members of a Neanderthal hunter-gatherer band gaze at the night sky and puzzle over a bright, red light that has recently appeared. Woolly mammoths and other Pleistocene Age animals are shown in the background. *(NASA; artist, Randii Oliver)*

western areas of Eurasia during the Pleistocene epoch in the time of the last Ice Age.

Neanderthals looked quite similar to human beings today—except that they had slightly pronounced foreheads, wider noses, and larger jaws. These prehistoric people were robust, short, and stocky. They were nomadic and survived by hunting and gathering. Neanderthals lived in the forest, on mountains, and within the plains regions of Ice Age Eurasia. Caves often provided them with a convenient shelter, but, being nomadic, they did not construct permanent individual habitats or villages. Plants were eaten seasonally, but about 90 percent of their diet consisted of meat. Pleistocene-epoch animals included the now extinct woolly mammoth, the bush-antlered deer, and the saber-toothed tiger.

Mimicking the behavior of the modern human, the Neanderthals wore garments made of leather and fur (primarily to protect them from the harsh environment), used fire, and made stone tools and weapons. They lived in small hunter-gatherer bands with a leadership structure, wore ornamental jewelry, and buried their dead ceremonially. Like most early peoples throughout the world, the Neanderthals looked up at the sky and probably made up stories about what they saw but could not physically explain.

Neanderthals appear to have become extinct about 30,000 to 35,000 years ago. Archaeologists and anthropologists now speculate that the Neanderthal was "out-competed" for resources and living space by a species of humans known as Cro-Magnons. In the clash of prehistoric cultures, primitive technologies, and levels of intellect, Neanderthals apparently lost and vanished from the face of Earth some 10,000 to 15,000 years after making contact with Cro-Magnons.

Is there a lesson from such early human behavior that is applicable to people in the space age? Perhaps. The later chapters of this book speculate about the consequences of contact with alien civilizations, especially those that are more advanced than 21st-century Earth. Should the modern descendants of Cro-Magnons risk "going the way of Neanderthals" by answering a "phone call" or opening a derelict space probe from an extraterrestrial civilization? The quest for life in the universe is filled with interesting circumstances and outcomes—some of which should prove to be very beneficial and uplifting for the human species; some of which may warrant caution and concern.

Prehistoric cave paintings (some as many as 30 millennia old) provide a lasting testament that early peoples engaged in stargazing and incorporated such astronomical observations in their cultures. In some ancient societies, the leading holy men would carve special astronomical symbols in stones (petroglyphs) at ancient ceremonial locations. Modern archaeologists and astronomers now examine and attempt to interpret these petroglyphs, as well as other objects that are uncovered in ancient ruins and that appear to have some astronomical significance. Did prehistoric peoples speculate about life in the universe beyond planet Earth? Probably not. Without written histories or records (hence the word *prehistoric*) to guide them, archaeologists, anthropologists, and other scientists can only speculate about what the objects in the sky meant to these early peoples.

For many ancient peoples, the daily motion of the Moon, the Sun, and the planets and the annual appearance of certain groups, or constellations, of stars served as natural calendars that helped regulate daily life. Since these celestial bodies were beyond physical reach or understanding, various mythologies emerged along with native astronomies. Within many early cultures, the sky became the home of the gods, and the Moon and the Sun were often deified.

While no anthropologist really knows what the earliest human beings thought about the sky, the culture of the Australian Aborigines—which has been passed down for more than 40,000 years through the use of legends, dances, and songs—gives collaborating anthropologists and astronomers a glimpse of how these early people interpreted the Sun, the Moon, and the stars. The Aboriginal culture is the world's oldest and most long-lived, and the Aboriginal view of the cosmos involves a close interrelationship between people, nature, and sky. Fundamental to their ancient culture is the concept of "the Dreaming"—a distant past when the spirit ancestors created the world. Aboriginal legends, dance, and songs express how in the distant past the spirit ancestors created the natural world and entwined people into a close relationship with the sky and with nature. Within the Aboriginal culture, the Sun is regarded as a woman. She awakes in her camp in the east each day and lights a torch that she then carries across the sky. In contrast, Aborigines consider the Moon as male, and because of the coincidental association of the lunar cycle with the female menstrual cycle, they linked the Moon with fertility and consequently gave it a great magical status. These ancient peoples also regarded a solar eclipse as the male Moon uniting with the female Sun.

The Egyptians and the Maya both used the alignment of structures to assist in astronomical observations and the construction of calendars. Modern astronomers have discovered that the Great Pyramid at Giza, Egypt, has a significant astronomical alignment, as do certain Maya structures such as those found at Uxmal, Yucatán, Mexico. Maya astronomers were particularly interested in times (called zenial passages) when the Sun crossed over certain latitudes in Central America. The Maya were also greatly interested in the planet Venus and treated the planet with as much importance as the Sun. These Mesoamerican native people had a good knowledge of astronomy and were able to calculate planetary movements and eclipses for millennia.

For the ancient Egyptians, Ra (also called Re) was regarded as the all-powerful sun god who created the world and sailed across the sky each day. As a sign of his power, an Egyptian pharaoh would use the title "son of Ra." Within Greek mythology, Apollo was the god who pulled the Sun across the sky, riding in his golden chariot with his twin sister Artemis (Diana in Roman mythology), the Moon goddess.

Based on available records and data, it is reasonable to conclude that from the dawn of human history until the start of the scientific revolution in the early 17th century, the heavens were regarded as an essentially unreachable realm—the abode of deities and, for some civilizations and religions, the place where a good, just person (or at least his or her conscious spirit) would go after physical life on Earth.

This 1964 Italian postage stamp honors the 400th anniversary of the birth of Galileo Galilei (February 15, 1564). A brilliant physicist, mathematician, and astronomer, Galileo Galilei founded the science of mechanics, promoted the scientific method, and fanned the flames of the scientific revolution by vigorously supporting the Copernican hypothesis—for which astronomical advocacy he was eventually found guilty of heresy and imprisoned (house arrest) for the remainder of his life. (Author)

So what happened that changed the human perception of heavens from an unreachable realm to a place to be visited? In other words, what encouraged people to begin to think about space travel and the possibility of life beyond Earth? The simple answer to this rather complex socio-technical question is the Copernican revolution (which started in about 1543) and the development of modern (observational) astronomy by Galileo Galilei, Johannes Kepler, and Sir Isaac Newton.

The first major step in this transition took place in 1609 when the Italian scientist Galileo Galilei (1564–1642) learned about a new optical instrument (a magnifying tube) that had just been invented in Holland. Within six months, Galileo devised his own version of the instrument. Then, in 1610, he turned this improved telescope to the heavens and started the age of telescopic astronomy. With his crude instrument, he made a series of astounding discoveries, including the existence of mountains on the Moon, many new stars, and the four major moons of Jupiter—now called the Galilean satellites in his honor. Galileo published these important discoveries in the book: *Sidereus Nuncius (Starry Messenger)*. The book stimulated both enthusiasm and anger. Galileo used the moons of Jupiter to prove that not all heavenly bodies revolve around Earth. This provided direct observational evidence for the Copernican model—a cosmological model that Galileo now began to endorse vigorously. The mountains on the Moon and the dark regions, which Galileo thought were oceans and seas and mistakenly called *mare*, suddenly made the Moon a physical *place* just like Earth.

If the Moon is indeed another world, and not some mysterious object in the sky, then inquisitive human beings might someday try to travel there. With the birth of optical astronomy in the early 17th century, not only was the arrival of the scientific revolution accelerated, but the embryonic notion of space travel and visiting other worlds suddenly acquired a touch of physical reality.

But seeing other worlds in a telescope was just the first step. The next critical step that helped to make the dream of space travel a reality was the development of a powerful machine that could not only lift objects off the surface of Earth but also operate in the vacuum of outer space. The modern rocket, as developed during World War II and vastly improved afterward in the cold-war era, became the enabling technology for human spaceflight—a pathway that would open many exciting future options for the human race.

However, even after the development of the modern rocket, there was one final step that was still needed to make human spaceflight a reality. One or more governments had to be willing to invest large quantities of money and engineer-

This is a scaled, composite image (scale factor: Each pixel equals 9.3 miles [15 km]) of the major members of the Jovian system—collected by NASA's *Galileo* spacecraft during various flyby encounters in 1996 and 1997. Included in the interesting family portrait is the edge of Jupiter with its Great Red Spot (GRS), as well as Jupiter's four largest moons, called the Galilean satellites. From top to bottom, the moons shown are Io, Europa, Ganymede, and Callisto. *(NASA/JPL)*

ing talent so that people could travel beyond Earth's atmosphere. From a historic perspective, the fierce geopolitical competition of the cold-war era between the United States and the former Soviet Union provided the necessary social stimulus. In an effort to dominate world political opinion in the 1960s, both governments decided to make enormous resource investments in the superpower "race into space."

The remainder of this chapter shows how each of these steps—the vision, the enabling hardware, and the political will—came together and made space travel a hallmark achievement of the human race in the latter portion of the 20th century. The apex of that technological achievement was the manned lunar landing missions of NASA's Apollo Project. The scientific search for life beyond Earth became an integral part of space exploration activities. Looking for alien life on other worlds left the realm of science fiction and grew into the exciting new field of science called exobiology (or astrobiology).

✧ Alien Creatures: Science Fiction or Fact?

From the start of the scientific revolution until the dawn of the space age (in 1957), the issue of whether alien life-forms inhabited other worlds in the solar system or possibly existed on planets around other stars resided primarily in the realm of science fiction. While a few scientists began to lay the foundations of exobiology in the 20th century by investigating conditions that could have started life in the primitive chemical soup of an ancient Earth, until the 1950s, most mainstream scientists politely skirted the topic of alien life. As discussed in this chapter and elsewhere in the book, there were, however, several notable exceptions, including Giordano Bruno (1548–1600), Svante August Arrhenius (1859–1927), Giovanni Virginio Schiaparelli (1835–1910), Percival Lowell (1855–1916), and Enrico Fermi (1901–54).

Science fiction is a form of fiction in which technical developments and scientific discoveries represent an important part of the plot or story background. Frequently, science fiction involves the prediction of future possibilities that are based on new scientific discoveries or technical breakthroughs. Some of the most popular science-fiction predictions that are waiting to happen are the discovery of alien life-forms, interstellar travel, contact with extraterrestrial civilizations, the development of exotic propulsion or communication devices that might permit people to break the speed-of-light barrier, travel forward or backward in time, and self-aware machines and robots. From the perspective of contemporary physics, some of these anticipated breakthroughs, such as superluminal travel, could prove impossible because of the physical laws and limits of the universe. Other developments, such as very intelligent machines, might take place faster than currently imagined.

According to the well-known writer Isaac Asimov (1920–92), one very important aspect of science fiction is not just its ability to predict a particular technical breakthrough but rather its ability to predict change itself through technology. Change plays a very important role in modern life. People who are responsible for societal planning must not only con-

GIORDANO BRUNO (1548–1600)

As a fiery philosopher and writer, the former Dominican monk Giordano Bruno managed to antagonize authorities throughout western Europe by adamantly supporting such politically sensitive and religiously unpopular (at the time) concepts as the heliocentric cosmology of Nicolaus Copernicus, the infinite size of the universe, and the existence of intelligent life on other worlds. While a member of the Dominican religious order, he had an abrasive personality and nurtured the tendency to voice his own, often extremely unpopular, opinions boldly to the annoyance of his fellow monks. To avoid prosecution for heresy, Bruno left the Dominican order when he was 28 years old and fled from Italy.

For the next 15 years, he traveled about Europe and continued to express his controversial thoughts, alienating local authorities wherever he spoke. His self-destructive, belligerent manner eventually brought him back to Italy—a fatal mistake that placed him in the legal grasp of the Roman Inquisition. After an eight-year-long trial, an uncompromising Bruno was finally convicted of heresy by the Roman Inquisition and burned to death at the stake in Campo de' Fiore square in Rome on February 17, 1600.

The often quoted passage in Bruno's controversial work *On the Infinite Universe and Worlds* (1584) states: "Innumerable suns exist; innumerable earths revolve around these suns in a manner similar to the way the seven planets revolve around our sun. Living beings inhabit these worlds." Some science historians use this passage to treat Bruno as a martyr to science and a hero of the scientific revolution. Other science historians point out that he was not an astronomer and that the basic theme of this controversial work dealt much more with his personal cosmology, based on pantheism, than with the emerging Copernican hypothesis. In fact, the ecclesiastical authorities in Rome did not formally ban the Copernican hypothesis until several years after Bruno's execution, so his alleged support for Copernican (heliocentric) cosmology may not have been the immediate cause of his death sentence.

Unlike the heresy trial of Galileo Galilei, which is well documented, the records stating the exact charges brought against Bruno during his lengthy trial for heresy in the last decade of the 16th century have been lost. So, more than four centuries after his fiery demise in a public execution, Giordano Bruno still remains a controversial personality in the history of science and in the advocacy of the existence of life beyond Earth.

sider how things are now but also how they will (or at least might) be in the upcoming decades. Gifted science-fiction writers, such as Jules Verne (1828–1905), Herbert George (H. G.) Wells (1866–1946), Isaac Asimov, and Sir Arthur C. Clarke (1917–   ), also serve society as skilled technical prophets who help many people to peek at tomorrow before it arrives.

For example, the famous French writer Jules Verne wrote *De la Terre à la Lune (From the Earth to the Moon)* in 1865, an account of a human voyage to the Moon from a Floridian launch site near a place that Verne called "Tampa Town." A little more than 100 years later, directly across the state from the modern city of Tampa, the once isolated regions of the east-central Florida coast shook to the mighty roar of a Saturn V rocket. The crew of NASA's *Apollo 11* mission had embarked from Earth, and people were to walk for the first time on the lunar surface. Verne's amazing stories stimulated interest in space travel and gave birth to the literary form of science fiction. He was followed by such other writers as H. G. Wells, who embellished their science-fiction stories with alien creatures in a variety of forms—often hostile to human beings and desirous of taking over Earth.

The basic roots of contemporary science fiction (books, motion pictures, and television shows) are found in scientific possibilities, not in magic or mysticism, but the boundary lines that separate future scientific reality, entertaining science fiction, and pure fantasy are incredibly blurred in an age where generational steps in technical progress occur in years versus decades, centuries, or even millennia. Arthur C. Clarke's often quoted third law of technical prophecy states: "Truly advanced technology is indistinguishable from magic." Ask anyone who was a child in the 1930s if they ever imagined people actually traveling into space, using personal electronic computers, or having robots assist surgeons during minimally invasive surgeries. Yet all such "fantastic ideas" are part of today's technical reality. The detection of extraterrestrial life, no matter how primitive, is one of those science-fiction themes or subgenres that could easily become science fact within the next few decades.

From the mid 19th century forward, actual scientific progress often stimulated science-fiction stories that reached just beyond the boundaries of the known or achievable. One example of this synergistic relationship occurred when the Italian astronomer Schiaparelli published a detailed map of the surface of Mars in 1877. He based this influential publication on a set of very precise astronomical observations of

The Italian astronomer Giovanni Virginio Schiaparelli (1835–1910) made careful observations of the planet Mars in the 1870s. As commemorated here on this 1974 Hungarian postage stamp, he produced a detailed map of the Red Planet's surface, including some straight markings, which he described as *"canali"*—the Italian word meaning "channels." Mistakenly translated into English as *canals*, Schiaparelli's work helped launch a frantic search for "canals" on Mars by some astronomers, who mistakenly considered such surface features as artifacts of an ancient, intelligent civilization. *(Author)*

Mars, which he performed earlier that year. An excellent observational astronomer, Schiaparelli dutifully described some straight markings as *canali*—meaning "channels" in Italian.

Unfortunately, when his description of such linear features on Mars was translated into English, the word *canali* improperly became *canals.* Some astronomers, most notably the wealthy U.S. astronomer Percival Lowell, completely misunderstood the true meaning of Schiaparelli's observations and launched a zealous telescopic search for the supposed canals that represented the handiwork of a neighboring intelligent alien civilization on the Red Planet. Schiaparelli never endorsed Lowell's extensive extrapolation of some of his best work in planetary astronomy. Yet, this erroneous interpretation of his *canali* in an otherwise excellent observational report about Mars is how most people remember the Italian astronomer. Schiaparelli's good astronomical work can also be viewed as starting an "alien stampede," in which the notion of intelligent extraterrestrial creatures, benign and malevolent, became more palpable and credible in both the science fact and fiction literature.

✧ Percival Lowell and the Canals of Mars

Late in the 19th century, the U.S. astronomer Percival Lowell established a private astronomical observatory (called the Lowell Observatory) near Flagstaff, Arizona, primarily to support his personal interest in Mars and his aggressive search for signs of an intelligent civilization there. Driving Lowell was his misinterpretation of Giovanni Schiaparelli's use of *canali* in an 1877 technical report in which the Italian astronomer discussed his telescopic observations of the Martian surface. Lowell took this report as early observational evidence of large, water-bearing canals built by intelligent beings. Lowell then wrote books, such as *Mars and Its Canals* (1906) and *Mars as the Abode of Life* (1908) to communicate his Martian civilization theory to the public.

While his nonscientific (but popular) interpretation of observed surface features on Mars proved quite inaccurate, his astronomical instincts were correct for another part of the solar system. Based on perturbations in the orbit of Neptune, Lowell predicted in 1905 the existence of a planet-sized, trans-Neptunian object. In 1930, the U.S. astronomer Clyde Tombaugh (1906–97), working at the Lowell Observatory, discovered Lowell's Planet X and called the tiny planet Pluto. The story of distant Pluto came full circle in August 2006 when members of the International Astronomical Union (IAU) (meeting in Prague, the Czech Republic) voted to demote Pluto from its traditional status as one of the nine major planets and placed the celestial object into a new class called a dwarf planet.

Percival Lowell was born on March 13, 1855, in Boston, Massachusetts, into an independently wealthy aristocratic family. His brother (Abbott Lowell) became president of Harvard University, and his sister (Amy) became an accomplished poet. Following graduation with honors from Harvard University in 1876, Lowell devoted his time to business and to traveling throughout the Far East. Based on his experiences between 1883 and 1895, Lowell published several books about the Far East.

He was not especially attracted to astronomy until later in life when he discovered an English translation of Schiaparelli's 1877 Mars observation report that included *canals* for the Italian word *canali*. (As originally intended by Schiaparelli, the word *canali* in Italian simply meant "channels.") Thus, in the early 1890s, Lowell became erroneously inspired by the thought of "canals" on Mars—that is, artificially constructed structures, the (supposed) presence of which he then extended to imply the existence of an advanced alien civilization. From this point on, Lowell decided to become an astronomer and then dedicated his time and wealth to a detailed study of Mars.

Lowell was unlike most other observational astronomers in that he was independently wealthy and already had a general idea of what he was searching for—namely, evidence of an advanced civilization of the Red Planet. He spared no expense to support this quest and obtained the assistance of several noted professional astronomers to help him find an excellent "seeing" site upon which to build this private observatory for the study of Mars. Constructed near Flagstaff, Arizona, the Lowell Observatory was opened in 1894 and housed a top-quality 24-inch (61-cm) refractor telescope that allowed Lowell to perform some excellent planetary astronomy. However, his observations tended to anticipate the things he reported, like oases and seasonal changes in vegetation. Unfortunately, the professional astronomers on his staff would label the same blurred features as simply indistinguishable natural markings. As Lowell more aggressively embellished his Mars observations, key staff members such as Andrew Ellicott Douglass (1867–1962) began to question Lowell's interpretation of these data. Perturbed by Douglass's scientific challenge, Lowell simply fired him in 1901 and then hired another professional astronomer, Vesto Melvin Slipher (1875–1969), to fill the vacancy.

In 1902, the Massachusetts Institute of Technology gave Lowell an appointment as a nonresident astronomer. He was definitely an accomplished observer but often could not resist the temptation to stretch greatly his interpretation of generally fuzzy and optically distorted surface features on Mars into observational evidence of the presence of artifacts from an advanced civilization. With such books as *Mars and Its Canals,* which was published in 1906, Lowell became popular with the general public who drew excitement from his speculative (but scientifically unproven) theory of an intelligent alien civilization on Mars, a civilization

A Jet Propulsion Lab (JPL, Pasadena, California) scientist carefully assembles the first primitive close-up image of Mars–taken by NASA's *Mariner 4* spacecraft as it flew past the Red Planet on July 14, 1965, at an altitude of 6,120 miles (9,846 km) above the surface. This historic flyby encounter provided scientists with their first glimpse of Mars at close range and put to rest all the popular speculations and myths, originating in the late 19th century, that the planet was home to an advanced civilization. The spacecraft took 22 television pictures, which covered about 1 percent of the Red Planet's surface and that revealed a vast, barren wasteland of craters, strewn about a rust-colored carpet of sand. Although the "engineered canals" reportedly observed by Percival Lowell in 1890 proved to be nothing more than an optical illusion, the *Mariner 4* images did suggest the possibility of ancient natural waterways in some regions of the planet. *(NASA/JPL)*

that was struggling to distribute water from the planet's polar regions with a series of elaborate giant canals. While most planetary astronomers shied away from such unfounded speculation, science-fiction writers flocked to Lowell's alien civilization hypothesis—a premise that survived in various forms until the dawn of the space age. On July 14, 1965, NASA's *Mariner 4* spacecraft flew past the Red Planet and returned images of its surface that shattered all previous speculations and romantic myths about a series of large canals that had been built by a race of ancient Martians.

Since the *Mariner 4* encounter with Mars, a large number of other robot spacecraft have studied Mars in great detail—both from orbit and on the surface. No cities, no canals, and no intelligent creatures have been found on the Red Planet. What has been discovered, however, is an interesting "halfway" world. Part of the Martian surface is ancient, like the surfaces of the Moon and Mercury, while part is more evolved and Earthlike. In this century, robot spacecraft and eventually human explorers will continue Lowell's quest for Martians—but this time, they will hunt for tiny microorganisms *possibly* living in sheltered biological niches or else frozen in time as fossilized evidence of ancient Martian life-forms that existed when the Red Planet was a milder, kinder, and wetter world.

While Lowell's quest for signs of intelligent life on Mars may have lacked scientific rigor by a considerable margin, his astronomical instincts about Planet X—his name for a suspected icy world lurking beyond the orbit of Neptune—proved correct. In 1905, Lowell began to make detailed studies of the subtle perturbations in Neptune's orbit and predicted the existence of a planet-sized trans-Neptunian object. He then initiated an almost decade-long telescopic search but failed to find this elusive object. In 1914, near the end of his life, he published the negative results of his search for Planet X and bequeathed the task to some future astronomer. Lowell died in Flagstaff, Arizona, on November 12, 1916.

✧ Herbert George (H. G.) Wells and Invaders from Mars

Another very influential science-fiction writer of the late 19th and early 20th century was Herbert George (H. G.) Wells. He inspired many future astronautical pioneers with his exciting fictional works that popularized the idea of space travel and life on other worlds. For example, in 1897, he wrote *The War of the Worlds*—the classic tale about extraterrestrial invaders from Mars.

Wells was born on September 21, 1866, in Bromley, Kent, England. In 1874, a childhood accident forced him to recuperate with a broken leg. The prolonged convalescence encouraged him to become an ardent reader, and this period of intensive self-learning served him well. He went on to become an accomplished author of both science fiction and more traditional novels.

He settled in London in 1891 and began to write extensively on educational matters. His career as a science-fiction writer started in 1895 with publication of the incredibly popular book *The Time Machine.* At the turn of the century, he focused his attention on space travel and the consequences of alien contact. Between 1897 and 1898, *The War of the Worlds* appeared as a magazine serial and then as a book. Wells followed this very popular space-invasion story with *The First Men in the Moon,* which appeared in 1901. Like Jules Verne, Wells did not link the rocket to space travel, but his stories did excite the imagination. *The War of the Worlds* was the classic tale of an invasion of Earth by creatures from space. In his original story, hostile Martians land in 19th-century England and prove to be unstoppable, conquering villains until tiny terrestrial microorganisms destroy them.

In writing this story, Wells was probably influenced by the then popular (but incorrect) assumption that allegedly observed Martian "canals" were artifacts of a dying civilization on the Red Planet. This was a very fashionable hypothesis in late 19th-century astronomy. As previously mentioned, the Martian canal craze had started quite innocently in 1877 when the Italian astronomer Giovanni Schiaparelli reported linear features that he observed on the surface of Mars as *canali*—the Italian word for *channels*. Schiaparelli's accurate astronomical observations became misinterpreted when translated as *canals* in English. Consequently, other notable astronomers such as the American Percival Lowell began to search enthusiastically for and soon "discover" other surface features on the Red Planet that resembled signs of an intelligent Martian civilization.

H. G. Wells cleverly solved (or more accurately ignored) the technical aspects of space travel in his 1901 novel, *The First Men in the Moon*. He did this by creating "cavorite"—a fictitious antigravity substance. His story inspired many young readers to think about space travel. However, space age missions to the Moon have now completely vanquished the delightful (though incorrect) products of this writer's fertile imagination, which included giant moon caves, a variety of lunar vegetation, and even bipedal creatures called Selenites.

The alien creatures whom Wells introduced in his science-fiction stories opened the floodgates of imagination and creativity. Soon, the science-fiction literature, motion pictures, and eventually television programs contained all manner of extraterrestrial life-forms, ranging from superintelligent species to aggressive alien insects, reptiles, and arachnids. This almost endless parade of fictional extraterrestrial creatures has included humanoid aliens; intelligent canine and feline species; aquatic aliens, plants, and fungi; conscious rocklike creatures; shape-shifting critters; parasites; and all manner of robots, androids, and cyborgs. Perhaps reflecting some of the human author or director's own extraterrestrial life chauvinisms, the vast majority of the fictional alien creatures resided on planets, although a few almost spiritlike, noncorporeal creatures, capable of living in outer space, have also appeared. (Extraterrestrial-life chauvinisms are discussed in chapter 3.)

However, in many of his other fictional works, Wells was often able to anticipate advances in technology correctly. This earned him the status of a technical prophet. For example, he foresaw the military use of the airplane in his 1908 work *The War in the Air* and foretold of the splitting of the atom in his 1914 novel *The World Set Free*.

Following his period of successful fantasy and science-fiction writing, Wells focused on social issues and the problems associated with emerging technologies. For example, in his 1933 novel, *The Shape of Things to Come*, he warned about the problems facing western civilization. In 1935, Alexander Korda produced a dramatic movie version of this futuristic tale. The movie closes with a memorable philosophical discussion on (technological) pathways for the human race. Sweeping an arm, as if to embrace the entire universe, one of the main characters asks his colleague: "Can it really be our destiny to conquer all this?" As the scene fades out, his companion replies: "The choice is simple. It is the whole universe or nothing. Which shall it be?"

The famous novelist and visionary died in London on August 13, 1946. He had lived through the horrors of two world wars and witnessed the emergence of many powerful new technologies, except space technology. His last book, *Mind at the End of Its Tether*, appeared in 1945. In this work, Wells expressed a growing pessimism about humanity's future prospects.

On October 30, 1938, the American actor and film director (George) Orson Welles (1915–85) produced a radio production of H. G. Wells's

ANDROIDS AND CYBORGS

An android is an anthropomorphic machine—that is, a robot with near-human form, features, and/or behavior. Although originating in science fiction, engineers and scientists now use the term *android* to describe robot systems that are being developed with advanced levels of machine intelligence and electromechanical mechanisms, so the machines can "act" like people. A future human-form field-geologist robot that was able to communicate with its human partner (as the team explored the surface of the Moon) by using a radio-frequency transmitter, as well as by turning its head and gesturing with its arms, would be an example of an android.

The term *cyborg* is a contraction of the expression: cybernetic organism. Cybernetics is the branch of information science that deals with the control of biological, mechanical, and/or electronic systems. While the term *cyborg* is quite common in contemporary science fiction—for example, the frightening "Borg collective" in the popular *Star Trek: The Next Generation* motion picture and television series—the concept was actually first proposed in the early 1960s by several scientists who were then exploring alternative ways of overcoming the harsh environment of space. The overall strategy that they suggested was simply to adapt a human being to space by developing appropriate technical devices that could be incorporated into an astronaut's body. With these implanted or embedded devices, astronauts would become cybernetic organisms, or cyborgs.

The War of the Worlds that brought many Americans to the state of near panic. The exceptionally realistic one-hour-long, live radio broadcast began just after eight o'clock that Sunday evening. Welles's fictitious radio show shocked listeners with the announcement of Martians landing at a farm in Grover Mills, New Jersey. It appears that listeners who lived closest to the make-believe landing site suffered the greatest amount of anxiety and fright. But the tale of this bogus invasion was broadcast to all regions of the United States, so people in other locations, relatively far from New Jersey, also experienced some level of panic and fright. Welles and a small band of actors and musicians working in the Mercury Theater of the Columbia Broadcasting System (CBS) in New York City staged the entire event. His clever use of special sound effects, realistic-sounding special bulletins, and even a parade of prestigious (though bogus) official speakers only enhanced the credibility and, therefore, the shock value of the "Martian invasion." To this day, the 1938 "Invasion from Mars" radio broadcast remains arguably one of the most famous delusions of the U.S. public that was accomplished by the entertainment media.

But the influence of H. G. Wells's famous tale of an alien invasion from Mars has extended far beyond the famous 1938 broadcast. In 1953, producer George Pal created an exciting, visually stunning film entitled *The War of the Worlds*. This motion picture was updated to the 1950s and

Instead of simply protecting an astronaut's body from the harsh space environment by enclosing the person in some type of spacesuit, space capsule, or artificial habitat (the technical approach actually chosen), the scientists, who advocated the cyborg approach boldly, asked: Why not create cybernetic organisms that could function in the harsh environment of space without special protective equipment? For a variety of technical, social, and political reasons, the proposed line of research quickly ended, but the term *cyborg* has survived.

Today, the term is usually applied to any human being (whether on Earth, under the sea, or in outer space) who uses a technology-based, body-enhancing device. For example, a person with a pacemaker, a hearing aid, or an artificial knee could be considered a cyborg. When a person straps on wearable, computer-interactive components, such as the special vision and glove devices that are used in a virtual reality system, that person has (in fact) become a temporary cyborg.

By further extension, the term *cyborg* is sometimes used to describe fictional artificial humans or very sophisticated robots with near-human (or superhuman) qualities. The Golem (a mythical clay creature in medieval Jewish folklore) and the Frankenstein monster (from Mary Shelley's classic 1818 novel *Frankenstein: The Modern Prometheus*) are examples of the former, while Arnold Schwarzenegger's portrayal of the superhuman terminator robot (in the *Terminator* motion-picture trilogy) is an example of the latter usage.

set in southern California. Actor Gene Barry played Wells's fictional hero, Dr. Clayton Forrester. In addition to the carnage being created in southern California, Pal's movie also depicted the large-scale, global invasion of Earth by the Martians. This pre–space age cinematic masterpiece set new standards for alien invasion films—including use of disintegration weapons and death rays by the alien creatures—and earned an Oscar for special effects.

The success of George Pal's invasion movie encouraged other movie-industry producers and directors to follow suit, following the advent of the space age. *Mars Attacks!* appeared in 1996 and was essentially a humorous spoof of all previous alien invasion movies. In 2005, director Steven Spielberg presented a contemporary adaptation of the H. G. Wells story, *The World of the Worlds.*

As an interesting historic note, this film was Spielberg's third major motion picture about extraterrestrials arriving here on Earth. In his previous alien movies (*E.T.—The Extra-Terrestrial* and *Close Encounters of the Third Kind*), the creatures from space were very intelligent and benevolent toward the human race. But Spielberg's third extraterrestrial film followed the more traditional hostile-invaders theme that had been pioneered by H. G. Wells. Once again, the invading alien creatures were powerful, malevolent, and had absolutely no intention of sparing the human race. Their demise was more a matter of human good fortune and serendipity rather than technical skill and force of arms. The consequences of contact between two technically mismatched, colliding civilizations has remained a frequent theme in modern science fiction. Scientists are also aware of a dilemma that they might be creating when they attempt to communicate with intelligent alien civilizations. Such interesting questions as "Who speaks for Earth?" and "Should human beings respond to an alien message?" are addressed later in the book.

✧ Other Influential Pre–Space Age Writers

The romantic pre–space age speculations about Venus resembling a tropical, younger Earth and about Mars containing a more-established advanced civilization were enhanced by several fiction writers in the 1930s and 1940s—especially the U.S. author Edgar Rice Burroughs (1875–1950) and the Irish novelist, scholar, and Christian apologeticist Clive Staples (C. S.) Lewis (1898–1963).

Burroughs is perhaps best known for his fictional stories involving Tarzan, the hero of the African jungle, but he also created interesting works in many genres. With respect to extraterrestrial life, his most famous series is the so-called Barsoom series, consisting of 10 science-fiction adventure books in which the hero, John Carter, plays a central role as a dashing swordsman from Earth who rescues and falls in love with Thuvia, the princess of Mars. This series was released from 1912 to 1964, with the last book (entitled *John Carter of Mars*) being published after Burroughs's death. Advances in space technology and science drained some of the adventure and excitement out of these science-fiction tales when NASA's *Mariner 4* spacecraft sent back images (in 1964) that showed "Barsoom" (Burroughs's fictional name for Mars) to be a barren, desolate world, completely devoid of artificial canals, ancient cities, or intelligent inhabitants.

Burroughs also wrote fictional tales about creatures on the Moon. *The Moon Maid* and *The Moon Men* both appeared in 1926. Starting in 1934, with the publication of his *Pirates of Venus,* Burroughs entertained his readers with adventures on the cloud-enshrouded planet. The last book, *The Wizards of Venus,* in his five-book Venus series was published posthumously in 1970—several years after such spacecraft as NASA's *Mariner 2* had collected data that showed that Venus was an uninhabitable inferno and not the romantic, tropical jungle world resembling a prehistoric Earth.

The Irish novelist C. S. Lewis is best known for his juvenile reader series *The Chronicles of Narnia,* but his science-fiction trilogy represents the first interesting literary attempt to address the issue of exotheology. Lewis cleverly used the genre of fantasy adventure that is set in outer space to discuss Christian morality. The first book in the trilogy, *Out of the Silent Planet,* was released in 1938, and the story takes place mostly on Mars. In *Perelandra* (published in 1943), the story shifts to Venus, which is presented as an oceanic paradise. In Lewis's story, Venus serves as a new Garden of Eden with a new Adam and Eve. The fictional hero, Elwin Ransom, must prevent the diabolical physicist Professor Weston from tempting Eve and causing a reenactment of the biblical Fall of Man on Perelandra. The third book in this trilogy, *That Hideous Strength,* was released in 1945. Most of this fantasy adventure story takes place on Earth and involves a titanic clash between good and evil.

Lewis himself was more interested in writing an interesting fantasy-adventure story that packed his intended moral message; he did not pay especially rigorous attention to the technical details of planetary astronomy. For his purpose, Mars (called Malacandra) was an inhabited and older world, while Venus (called Perelandra) was an inhabited, younger world, resembling a tropical paradise. Within Lewis's fantasy lexicon for the solar system (called the field of Arbol), Earth is known as Thulcandra (meaning "the silent planet") and the Moon is called Sulva.

Several decades into the space age, it is quite difficult to assess properly the impact that the works of Burroughs and Lewis may have had on young minds in the 1940s and 1950s—the children who grew up and became the first generation of aerospace engineers, space scientists, and astronauts. But one impact is certain: For millions of young readers, the notion of intelligent life and alien creatures on other worlds became a possibility that was now a little less shocking and perhaps a bit more interesting.

✦ The Space Age View of Mars

Mars is the fourth major planet from the Sun and has an equatorial diameter of about 4,222 miles (6,794 km). Throughout human history, the Red Planet has been at the center of astronomical interest. The ancient Babylonians followed the motions of this wandering red light across the night sky and named it after Nergal, their god of war. In time, the Romans, also honoring their own god of war, gave the planet its present name.

The presence of an atmosphere, polar caps, and changing patterns of light and dark on the surface caused many pre–space age astronomers and scientists to consider Mars an "Earth-like planet"—the possible abode of extraterrestrial life. In fact, this unsupported but very popular belief was so widespread before the space age that when actor Orson Welles broadcast his *The War of the Worlds* radio drama in 1938, enough people believed the radio report of invading Martians so as to create a near panic in some areas of the United States. The flying saucer, or unidentified flying object (UFO), phenomenon that started shortly after World War II is another manifestation of how deeply ingrained that the popular notion of extraterrestrial visitors coming to Earth is in modern human society. (UFO phenomena are discussed in chapter 11.) Space technology has allowed scientists to put to rest some of the most imaginative visions of intelligent life on Mars.

Starting in 1964 with the successful flyby of NASA's *Mariner 4* spacecraft, a parade of sophisticated robot spacecraft—flybys, orbiters, and landers—have shattered all prevailing romantic myths of an ancient race of intelligent Martians struggling to bring water to the more productive regions of a dying world. *Mariner 4* flew by Mars on July 14, 1965, and returned images of the planet's surface. The closest encounter distance for this mission was 6,120 miles (9,846 km) from the Martian surface. *Mariner 6* passed within 2,132 miles (3,431 km) of the Martian surface on July 31, 1969. The spacecraft's instruments took images of the planet's surface and measured ultraviolet and infrared emissions of the Martian atmosphere. The *Mariner 7* spacecraft was identical to the *Mariner 6* spacecraft. *Mariner 7* passed within 2,130 miles (3,430 km) of Mars on August 5, 1969, and acquired images of the planet's surface. Emissions of the Martian atmosphere also were measured.

This 1976 U.S. postage stamp commemorated U.S. space-exploration achievements, especially NASA's Viking Project mission to Mars—the first sophisticated scientific attempt to find life on another planet. *(Author)*

Mariner 9 arrived at and orbited Mars on November 14, 1971. This spacecraft gathered data on the atmospheric composition, density, pressure, and temperature of the Martian atmosphere as well as performing studies of the Martian surface. After depleting its supply of attitude-control gas, the spacecraft was turned off on October 27, 1972. Consequently, spacecraft-derived data began to show that the Red Planet is actually a "halfway" world: Part of the Martian surface is ancient, like the surfaces of the Moon and Mercury, while part is more evolved and Earthlike.

But the question of whether life emerged on Mars in ancient times, when water possibly flowed on its surface, remained tantalizingly unresolved. To help answer that question, NASA developed the Viking Project—at the time, the most advanced and sophisticated combination of lander and orbiter robot spacecraft ever developed.

In August and September 1975, two Viking spacecraft were launched on a mission to help answer the question: Is there life on Mars? Each Viking spacecraft consisted of an orbiter and a lander. While scientists did not expect these spacecraft to discover Martian cities bustling with intelligent life, the exobiology experiments on the lander were designed to find evidence of primitive life-forms, past or present. (How spacecraft, such as the *Viking 1* and *2* landers, search for alien life is discussed in chapter 3.) Unfortunately, the results sent back by the two robot landers proved teasingly inconclusive.

The Viking Project was the first mission to soft-land a robot spacecraft successfully on another planet (excluding Earth's Moon.) All four Viking spacecraft (two orbiters and two landers) exceeded by considerable margins their design goal lifetime of 90 days. The spacecraft were launched in 1975 and began to operate around or on the Red Planet in 1976. When the *Viking 1* lander touched down on Chryse Planitia (the Plains of Gold) on July 20, 1976, it found a bleak landscape. Many ancient channels that could have once contained flowing surface water characterize this region of Mars.

Several weeks later, its twin, the *Viking 2* lander, set down on Utopia Planitia (the Plains of Utopia) and discovered a more gentle, rolling landscape. One by one, these robot explorers finished their highly successful visits to Mars. The *Viking 2* orbiter spacecraft ceased operation in July 1978; the *Viking 2* lander fell silent in April 1980; the *Viking 1* orbiter managed at least partial operation until August 1980; the *Viking 1* lander spacecraft made its final transmission on November 11, 1982. NASA officially ended the Viking Project's mission to Mars on May 21, 1983.

As a result of these interplanetary missions, scientists now know that Martian weather changes very little. For example, the highest atmosphere temperature recorded by either Viking lander was -5.8°F (-21°C) midsummer at the *Viking 1* site, while the lowest recorded temperature was -191°F (-124°C) at the more northerly *Viking 2* site during winter.

The atmosphere of Mars was found to be primarily carbon dioxide (CO_2). Nitrogen, argon, and oxygen are present in small percentages, along with trace amounts of neon, xenon, and krypton. The Martian atmosphere contains only a wisp of water (about 1/1,000th as much as found in Earth's atmosphere). But even this tiny amount can condense and form clouds that ride high in the Martian atmosphere or form patches of morning fog

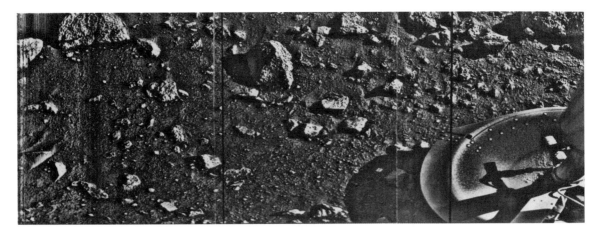

This image is this first photograph ever taken from the surface of Mars. The image was captured by NASA's *Viking 1* lander spacecraft shortly after it touched down on the surface of the Red Planet on July 20, 1976. Part of the spacecraft's number-two footpad is visible in the lower right corner; the sand and the dust that appear in the center of the footpad probably had been deposited during landing. *(NASA/JPL)*

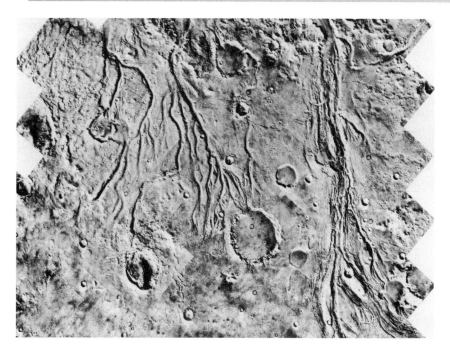

This mosaic image of the Martian surface was assembled from pictures collected by the *Viking 1* orbiter spacecraft between August 4 and 9, 1976. The area shown is centered at 17 degrees North, 55 degrees West—a region just to the west of the *Viking 1* lander site in Chryse Planitia. The channels that appear in this mosaic image are suggestive of massive ancient flood of waters from the Lunae Planum region, across this intervening cratered terrain, and into the general region of the *Viking 1* lander site. *(NASA/JPL)*

in valleys. There is also evidence that Mars had a much denser atmosphere in the past—one capable of permitting liquid water to flow on the planet's surface. Physical features resembling riverbeds, canyons and gorges, shorelines, and even islands hint that large rivers and maybe even small seas once existed on the Red Planet. As discussed in chapter 4, the current NASA strategy for exploring the Red Planet for signs of life—extinct or still existent in some sheltered subsurface biological niche—centers around the idea of "following the water."

✧ Martian Meteorites

NASA scientists now believe that more than 30 unusual meteorites are actually pieces of Mars that were blasted off the Red Planet by meteoroid impact collisions. These Martian meteorites were previously called

SNC meteorites, after the first three types of samples discovered (namely: Shergotty, Nakhla, and Chassigny). The Chassigny meteorite was discovered in Chassigny, France, on October 3, 1815. It establishes the name of the chassignite-type subgroup of the SNC meteorites. Similarly, the Shergotty meteorite fell on Shergotty, India, on August 25, 1865, and provides the name of the shergottite-type subgroup of SNC meteorites. Finally, the Nakhla meteorite was found in Nakhla, Egypt, on June 28, 1911, and establishes the name for the nakhlite-type subgroup of SNC meteorites. Another member of the Martian meteorite family, the 40-pound (18-kg) Zagami meteorite fell to Earth on October 3, 1962, near Katsina, Nigeria.

Martian meteorites have been found on every continent, except Australia. Unlike some meteorites that landed elsewhere and became unprotected commercial items for collectors, the samples found in Antarctica are collected and controlled by government organizations and curated for research by professional scientists.

All of the Martian meteorites are igneous rocks that have crystallized from molten lava in the crust of the parent planetary body. The meteorites that have been linked to Mars and that have been discovered on Earth so far represent five different types of igneous rocks, ranging from simple plagioclase-pyroxene basalts to almost monomineralic cumulates of pyroxene or olivine.

The only natural process capable of launching Martian rocks to Earth is meteoroid impact. To be ejected from Mars, a rock must reach a velocity of at least 3.1 miles per second (5 km/s)—the escape velocity for Mars. During a large meteoroid impact on the surface of Mars, the kinetic energy of the incoming cosmic impactor causes shock deformation, heating, melting, and vaporizing, as well as crater excavation and ejection of target material. The impact and the shock environment of such a collision provide scientists with an explanation as to why the Martian meteorites are all igneous rocks. Martian sedimentary rocks and soil would not be consolidated sufficiently to survive the impact as intact rocks and then wander through space for millions of years and eventually land on Earth as meteorites.

One particular Martian meteorite, called ALH84001, has stimulated a great deal of interest in the possibility of life on Mars. In August 1996, a NASA research team at the Johnson Space Center (JSC) announced that they had found four lines of evidence in ALH84001 that "strongly suggests primitive life may have existed on Mars more than 3.6 billion years ago." The NASA research team found the first organic molecules thought to be of Martian origin; several mineral features characteristic of biological activity; and possibly microscopic fossils of primitive, bacterialike organisms inside an ancient Martian rock that fell to Earth as a meteorite. While the NASA research team did not claim that they had conclusively proved

ALH84001,0

This 4.5-billion-year-old rock, labeled meteorite ALH84001, is believed at one time to have been a part of Mars. Some scientists speculate that this "Martian meteorite" may also contain fossil evidence that primitive life once existed on the Red Planet— perhaps more than 3.6 billion years ago. The rock is a portion of a meteorite that was apparently dislodged from Mars by a huge impact about 16 million years ago, slowly wandered through space, and then fell to Earth in Antarctica some 13,000 years ago. The meteorite was found in Antarctica's Allan Hills ice field in 1984. Today, the interesting rock is being preserved for scientific study at the Johnson Space Center's Meteorite Processing Laboratory. *(NASA/JSC)*

that life existed on Mars some 3.6 billion years ago, they did believe that "they have found quite reasonable evidence of past life on Mars."

Martian meteorite ALH84001 is a 4.2-pound (1.9-kg), potato-sized igneous rock that has been age-dated to about 4.5 billion years—the period scientists believe when the planet Mars formed. This rock is thought to have originated underneath the Martian surface and to have been fractured extensively by impacts as meteorites bombarded the planet during the early history of the solar system. Between 3.6 and 4.0 billion years ago, Mars is thought to have been a warmer and wetter world. Martian water could have penetrated fractures in the subsurface rock, possibly forming an underground water system. Since the water was saturated with carbon dioxide from the Martian atmosphere, carbonate materials were deposited in the fractures.

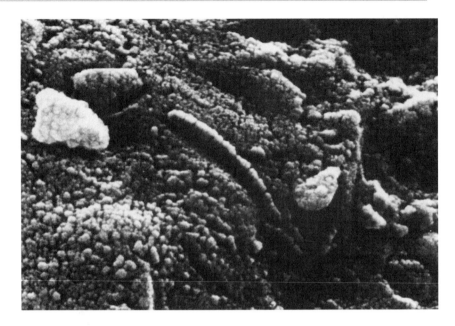

This high-resolution scanning electron microscope image shows an unusual tubelike structure (less than 1/100th the width of a human hair in size) found in meteorite ALH84001—a meteorite believed to be of Martian origin. Although this structure is not part of the research reported by NASA exobiologists in summer 1996, the object is located in a similar carbonate glob in the meteorite. This structure remains the subject of scientific investigations and of debates concerning whether or not it is fossil evidence of primitive life on Mars some 3.6 billion years ago. *(NASA/JSC)*

The NASA research team estimated that this rock from Mars entered Earth's atmosphere about 13,000 years ago and fell in Antarctica as a meteorite. ALH84001 was discovered in 1984 in the Allan Hills ice field of Antarctica by an annual expedition of the National Science Foundation's Antarctic Meteorite Program. It was preserved for study at the NASA JSC Meteorite Processing Laboratory, but its possible Martian origin was not fully recognized until 1993. It is the oldest of the Martian meteorites yet discovered.

Other scientists have used data from the *Mars Global Surveyor* and *Mars Odyssey 2001* spacecraft in an attempt to find a reasonable candidate location on Mars from which this very controversial ALH84001 meteorite may have originated. In 2005, they reported that ALH84001 may have come from the Eos Chasma, a branch of the very long and large Valles Marineris canyon system. The most likely candidate area is a 12.4-mile (20-km) diameter crater in a lobate flow region—the type of geologic phenomenon that occurs when a high-velocity impactor strikes a fluid-rich soil. The site has 2.5-mile (4-km)-high cliffs, bordering the canyon and exposing rocks from various geologic ages in the history of Mars. The scientists further suggest that ALH84001 may have first formed in this region deep beneath the Martian surface and then was later transported to the shallow depth from which an ancient, high-speed impactor launched the famous rock into space.

However, the birthplace on Mars of ALH84001 is still a matter of much conjecture, as is the hypothesis that it contains fossilized evidence suggesting there once was primitive life on Mars. In any event, Eos Chasma appears to be an interesting region from which to sample rocks from a wide range of ages in the history of Mars and could become a prime candidate target for a future robotic or human landing mission.

More than a decade after NASA scientists at the Johnson Space Center made their startling announcement about ALH84001, the question of life on ancient Mars remains a wide-open question. The only aspect about this controversial rock about which almost all scientists agree is that it came from Mars and that it is older than any known rock on Earth. ALH84001 was volcanically formed; was determined to be incredibly old—almost as old as the solar system itself; has the appropriate elemental fingerprints for Mars—consistent with the chemistry data recorded by the *Viking 1* and *2* landers in 1976; contains materials that are rare on Earth; and, finally, appears to have been exposed to flowing water at some point before it was hurled off the Red Planet. Whether this interesting Martian meteorite actually contains evidence that is conclusively suggestive of ancient biogenic activity on Mars remains an open, bitterly contested scientific debate.

Some scientists currently hold the position that neither ALH84001 nor any other Martian meteorite will lead them to establish an irrefutable, scientific conclusion about whether life actually existed on Mars. What these exobiologists want, and may someday obtain, are pristine samples of Martian rock and soil, collected and returned to Earth for scientific investigation under controlled conditions. (Chapter 4 discusses a Mars sample return mission.) For now, AHL84001 has created a renaissance in the search for life beyond Earth and triggered a wave of new exploration of the Red Planet.

Life in the Universe

A ny search for life in the universe requires that scientists develop and agree upon a basic definition of what life is. For example, according to contemporary exobiologists and biophysicists, life (in general) can be defined as a living system that exhibits the following three basic characteristics: First, a living system is structured and contains information; second, an entity that is considered alive must be able to replicate itself; third, a living system experiences few random changes in its information package—which random changes when they do occur enable the living system to evolve in a Darwinian context (that is, survival of the fittest).

The history of life in the universe can be explored in the context of a grand, synthesizing scenario called cosmic evolution. This sweeping scenario links the development of galaxies, stars, planets, life, intelligence, and

The scenario of cosmic evolution provides a grand synthesis for the long series of alterations of matter and energy that have given rise to the Milky Way galaxy, the Sun, planet Earth, and human beings. *(NASA [from the work of Eric J. Chaisson])*

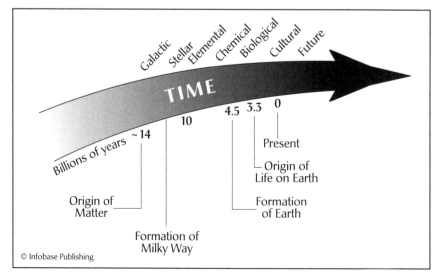

ANTHROPIC PRINCIPLE

The anthropic principle is an interesting, though highly controversial hypothesis in modern cosmology, which suggests that the universe evolved after the big bang in just the right way so that life, especially intelligent life, could develop. The proponents of this hypothesis contend that the fundamental physical constants of the universe (such as the speed of light [c] and the Planck constant [h]) actually support the existence of life and (eventually) the emergence of conscious intelligence—including, of course, human beings. In fact, the advocates of the anthropic principle are quick to point out, if the universe was not so suitable for life, people would simply not be around today inquiring about why things in the universe are so finely tuned and adjusted. The supporters of this hypothesis further suggest that with just a slight change in the value of any of these fundamental physical constants, the universe would have evolved very differently after the big bang.

The hypothesis that there could exist many different universes—each with different values of the physical constants—is sometimes referred to as the *strong anthropic principle.* Among these numerous possible universes, human beings emerged in the one particular universe—the one that contained just the right physical constants to permit carbon atoms to form and eventually to serve as the building blocks of living systems.

In contrast, the *weak anthropic principle* gives no operational meaning or significance to the concept of other universes; recognizes and accepts the importance of the physical constants that are found in and define the present universe; and seeks to interpret the significance of these physical constants with respect to the rise of life and consciousness. For example, soon after the big bang, only hydrogen and helium existed. Physicists who support the weak anthropic principle try to relate the creation of carbon and the other elements necessary for life within the context of a universe governed by the existing physical constants.

In exploring the implications of the anthropic principle, scientists sometimes like to play "what-if" mind games or resort to gedanken (thought) experiments. For example, if the force of gravitation (as manifested by the gravitational constant G) was weaker than it is, the expansion of matter after the big bang would have been much more rapid, and the development of stars, planets, and galaxies from extremely sparse (nonaccreting) nebular materials could not have occurred. No stars, no planets, no development of carbon-based life—as scientists currently know it. If, on the other hand, the force of gravitation was stronger than it is, the expansion of primordial material would have been sluggish and retarded, encouraging a total gravitational collapse (that is, the big crunch) long before the development of stars and planets. Again, no stars, no planets, no life.

Opponents of the anthropic principle (weak or strong) suggest that the values of the fundamental physical constants are just a coincidence. Within their line of reasoning, intelligence is not regarded as an inevitable byproduct of the evolution of matter and energy in a universe governed by the current physical constants. Rather, the evolution of biological systems with intelligence is only one of a number of possible outcomes.

The anthropic principle remains a lively topic for debate and speculation within the scientific community. Does the presence of the human species on Earth today represent the

(continues)

(continued)

inevitable byproduct of an evolving universe defined by its physical constants in just such a way so as to bring forth intelligent human consciousness? Does the anthropic principle apply elsewhere in the universe, or are human beings the most advanced, intelligent species thus far evolved in the entire universe? Until this controversial hypothesis can actually be tested, however, it must remain—by the rules of the scientific method—outside of the mainstream of demonstrable scientific principles.

technology and then speculates on where the ever-increasing complexity of matter is leading. The emergence of conscious matter (especially intelligent creatures), the subsequent ability of a portion of the universe to reflect upon itself, and the role and destiny of these intelligent creatures are topics often associated with contemporary discussions of cosmic evolution. One interesting speculation is the *anthropic principle*—namely, the hypothesis that the universe was designed for life, especially the emergence of human life.

The cosmic evolution scenario is not without scientific basis. The occurrence of organic compounds in interstellar clouds, in the atmospheres of the giant planets of the outer solar system, and in comets and meteorites suggests the existence of a chain of astrophysical processes that

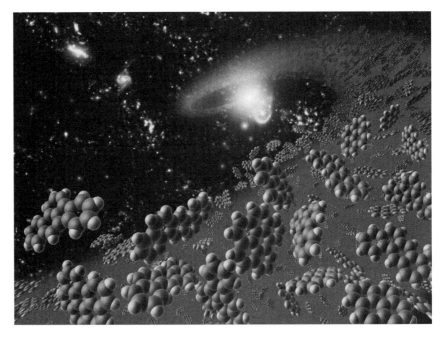

This artist's rendering symbolically represents complex organic molecules, known as polycyclic aromatic hydrocarbons, observed in the early universe. Scientists consider these large molecules, comprised of carbon and hydrogen, among the building blocks of life. *(NASA/JPL/Caltech)*

HOW OLD IS PLANET EARTH?

Planet Earth is very old, about 4.5 billion years or more, according to recent scientific estimates. Most of the evidence for an ancient Earth is contained in the rocks that form the planet's crust. The rock layers themselves, like pages in a long and complicated history, record the surface-shaping events of the past, and buried within them are traces of life—that is, the plants and animals that evolved from the ancient organic structures that existed perhaps three billion years ago. Also contained in once molten rocks are radioactive elements whose isotopes provide Earth with an atomic clock. Within these rocks, parent isotopes decay at a predictable rate to form daughter isotopes. By determining the relative amounts of parent and daughter isotopes, scientists have determined the age of these rocks. Therefore, the results of studies of rock layers (*stratigraphy*) and of fossils (*paleontology*), coupled with the ages of certain rocks as measured by atomic clocks (*geochronology*), provide scientific evidence that humans' home planet is a very old place.

Up until now, scientists have not found a way to determine the exact age of Earth directly from terrestrial rocks because our planet's oldest rocks have been recycled and destroyed by the process of plate tectonics. If there are any of Earth's primordial rocks left in their original state, scientists have not yet found them. Nevertheless, scientists have been able to determine the probable age of the solar system and to calculate an age for Earth by assuming that Earth and the rest of the solid bodies in the solar system formed at the same time and are, therefore, of the same age.

The ages of Earth and Moon rocks and of meteorites are measured by the decay of long-lived radioactive isotopes of elements that occur naturally in rocks and minerals and that decay with half-lives of 700 million to more than 100 billion years into stable isotopes of other elements. Scientists use these dating techniques, which are firmly grounded in physics and are known collectively as radiometric dating, to measure the last time that the rock being dated was either melted or disturbed sufficiently to rehomogenize its radioactive elements. Ancient rocks that exceed 3.5 billion years in age are found on all of Earth's continents. The oldest terrestrial rocks found so far are the Acasta Gneisses in northwestern Canada near Great Slave Lake (4.03 billion years old) and the Isua Supracrustal rocks in West Greenland (about 3.7 to 3.8 billion years old), but well-studied rocks nearly as old are also found in the Minnesota River Valley and northern Michigan (some 3.5–3.7 billion years old), in Swaziland (about 3.4–3.5 billion years old), and in Western Australia (approximately 3.4–3.6 billion years old). Scientists have dated these ancient rocks using a number of radiometric dating methods, and the consistency of the results gives scientists confidence that the estimated ages are correct to within a few percent.

An interesting feature of these ancient rocks is that they are not from any sort of "primordial crust" but are lava flows and sediments deposited in shallow water—an indication that Earth history began well before these rocks were deposited. In Western Australia, single zircon crystals found in younger sedimentary rocks have radiometric ages of as much as 4.3 billion years, making these tiny crystals the oldest materials to be found on Earth so far. Scientists have not yet found the source rocks for these zircon crystals.

(continues)

(continued)

The ages measured for Earth's oldest rocks and oldest crystals show that humans' home planet is at least 4.3 billion years in age, but do not reveal the exact age of Earth's formation. The best age for Earth is currently estimated as 4.54 billion years. This value is based on old, presumed single-stage leads (Pb) coupled with the Pb ratios in troilite from iron meteorites, specifically the Canyon Diablo meteorite. In addition, mineral grains (zircon) with uranium-lead (U-Pb) ages of 4.4 billion years have recently been reported from sedimentary rocks found in west-central Australia.

The Moon is a more primitive planetary body than Earth because it has not been disturbed by plate tectonics. Therefore, some of the Moon's more ancient rocks are more plentiful. The six American Apollo Project human missions and three Russian Luna robotic spacecraft missions returned only a small number of lunar rocks to Earth. The returned lunar rocks vary greatly in age—an indication of their different ages of formation and their subsequent geologic histories. The oldest dated Moon rocks, however, have ages between 4.4 and 4.5 billion years and provide a minimum age for the formation of Earth's nearest planetary neighbor.

Thousands of meteorites, which are fragments of asteroids that fell on Earth, have been recovered. These primitive extraterrestrial objects provide the best ages for the time of formation of the solar system. There are more than 70 meteorites (of different types) whose ages have been measured using radiometric-dating techniques. The results show that the meteorites, and, by extrapolation, the solar system, formed between 4.53 and 4.58 billion years ago. The best age for Earth comes not from dating individual rocks but by considering Earth and meteorites as part of the same evolving system in which the isotopic composition of lead, specifically the ratio of lead-207 to lead-206, changes over time, owing to the decay of radioactive uranium-235 and uranium-238, respectively. Scientists have used this approach to determine the time required for the isotopes in Earth's oldest lead ores, of which there are only a few, to evolve from its primordial composition, as measured in uranium-free phases of iron meteorites, to its compositions at the time that these lead ores separated from their mantle reservoirs.

According to scientists at the U.S. Geological Survey (USGS), these calculations result in an age for Earth and meteorites, and therefore the solar system, of 4.54 billion years with an uncertainty of less than 1 percent. To be precise, this age represents the last time that lead isotopes were homogeneous throughout the inner solar system and the last time that lead and uranium were incorporated into the solid bodies of the solar system. The age of 4.54 billion years found for the solar system and Earth is consistent with current estimates of 11 to 13 billion years for the age of the Milky Way galaxy (based on the stage of evolution of globular cluster stars) or the estimated age of 10 to 15 billion years (based on the recession rate of distant galaxies).

links the chemistry of interstellar clouds with the prebiotic evolution of organic matter in the solar system and on early Earth. There is also compelling evidence that cellular life existed on Earth some 3.56 billion years ago (3.56 Gy). This implies that the cellular ancestors of contemporary terrestrial life emerged rather quickly (on a geologic time scale). These ancient creatures may have also survived the effects of large impacts from comets and asteroids in those ancient, chaotic times when the solar system was evolving.

The accompanying figure summarizes some of the factors that scientists now believe are important in the evolution of complex life. These factors include (A) endogenous factors stemming from physical-chemical properties of Earth and those of *eukaryotic* organisms; (B) factors associated with properties of the Sun and of Earth's position with respect to the Sun; (C) factors originating within the solar system, including Earth as a representative planet; and (D) factors originating in space far from humans' solar system.

The word *eukaryotic* refers to cells whose internal construction is complex, consisting of organelles (e.g., nucleus, mitochondria, etc.), chromosomes, and other structures. All higher terrestrial organisms are built of eukaryotic cells, as are many single-celled organisms (called protists). The evolution of complex life apparently had to await the evolution of eukaryotic cells—an event that is believed to have occurred on Earth about one billion years ago. A *eukaryote* is an organism built of eukaryotic cells.

And where does all this lead to in the cosmic evolution scenario? The reader should first recognize that all living things are extremely interesting pieces of matter. Life forms that have achieved intelligence and have developed technology are especially interesting and valuable in the cosmic evolution of the universe. Intelligent creatures with technology, including human beings here on Earth, can exercise conscious control over matter in progressively more effective ways as the level of their technology grows. Ancient cave dwellers used fire to provide light and warmth. Modern humans harness solar energy, control falling water, and split atomic nuclei to provide energy for light, warmth, industry, and entertainment. People later in this century will most likely "join atomic nuclei" (controlled fusion) to provide energy for light, warmth, industrial applications, and entertainment here

These factors are considered important in the evolution of complex life. *(NASA)*

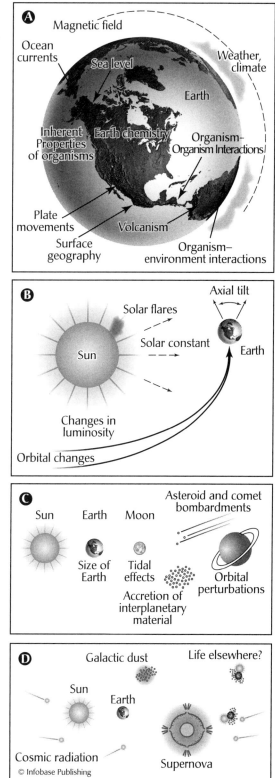

on Earth, as well as for interplanetary power and propulsion systems for emerging human settlements on the Moon and Mars. The trend should be obvious.

Some scientists, while contemplating the ultimate fate of the universe, have boldly suggested an interesting destiny for intelligent species throughout the universe. If such (postulated) intelligent alien creatures throughout the Milky Way galaxy can learn to live with the awesome technical powers of their advanced civilizations, then it may be the destiny of advanced intelligent life (including hopefully humans) to share information and to cooperate in the exercise of (benevolent) control over all the matter and energy within the universe.

According to modern scientific theory, living organisms arose naturally on the primitive Earth through a lengthy process of chemical evolution of organic matter. This process began with the synthesis of simple organic compounds from inorganic precursors in the atmosphere; continued in the oceans, where these compounds were transformed into increasingly more complex organic substances; and then culminated with the emergence of organic microstructures that had the capability of rudimentary self-replication and other biochemical functions.

Human interest in the origins of life extends back deep into antiquity. Throughout history, each society's creation myth seemed to reflect that particular people's view of the extent of the universe and their place within it. Today, in the space age, the scope of those early perceptions has expanded well beyond the reaches of humans' solar system to other star systems, to the vast interstellar clouds, and to numerous galaxies that populate the seemingly limitless expanse of outer space. Just as the concept of biological evolution implies that all living organisms have arisen by divergence from a common ancestry, so too the concept of cosmic evolution implies that all matter in humans' solar system has a common origin. Following this line of reasoning, scientists now postulate that life may be viewed as the product of countless changes in the form of primordial stellar matter—changes brought about by the interactive processes of astrophysical, cosmochemical, geological, and biological evolution.

If scientists use the even larger context of cosmic evolution, they can further conclude that the chain of events, which led to the origins of life here on Earth, extends well beyond planetary history: to the origin of the solar system itself, to processes occurring in ancient interstellar clouds that spawned stars like the Sun, and ultimately to the very birth within these stars (through nucleosynthesis and other processes) of the elements that make up living organisms—the *biogenic elements*. The biogenic elements are those that are generally judged to be essential for all living systems. Scientists currently place primary emphasis on the elements hydrogen (H), carbon (C), nitrogen (N), oxygen (O), sulfur (S) and phosphorous

(P). The compounds of major interest are those normally associated with water and with organic chemistry, in which carbon is bonded to itself or to other biogenic elements. The essentially universal presence of these compounds throughout interstellar space gives exobiologists the scientific basis for forming the important contemporary hypothesis that the *origin of life is inevitable throughout the cosmos wherever these compounds occur and suitable planetary conditions exist*. Present-day understanding of life on Earth leads scientists to the conclusion that life originates on planets and that the overall process of biological evolution is subject to the often chaotic processes that are associated with planetary and solar-system evolution—for example, the random impact of a comet on a planetary body or the unpredictable breakup of a small moon.

Scientists now use four major epochs in describing the evolution of living systems and their chemical precursors. These are:

1. The cosmic evolution of biogenic compounds—an extended period corresponding to the growth in complexity of the biogenic elements from nucleosynthesis in stars to interstellar molecules to organic compounds in comets and asteroids.
2. Prebiotic evolution—a period corresponding to the development (in planetary environments) of the chemistry of life from simple components of atmospheres, oceans, and crustal rocks to complex chemical precursors to initial cellular life-forms.
3. The early evolution of life—a period of biological evolution from the first living organisms to the development of multicellular species.
4. The evolution of advanced life—a period characterized by the emergence of progressively more advanced life-forms, climaxing perhaps with the development of intelligent beings capable of communicating, using technology and exploring and understanding the universe within which they live.

As scientists unravel the details of the intriguing process for the chemical evolution of terrestrial life, they should also ask themselves another very intriguing question: If it happened here, did it or could it happen elsewhere? In other words, what are the prospects for finding extraterrestrial life—in this solar system or perhaps on Earthlike planets around distant stars?

According to the *principle of mediocrity* (a concept frequently invoked by exobiologists), there is nothing "special" about the solar system or planet Earth. Within this speculation, therefore, if similar conditions have existed or are now present on "suitable" planets around alien suns, the chemical evolution of life will also occur.

Contemporary planetary formation theory strongly suggests that objects similar in mass and composition to Earth may exist in many

PRINCIPLE OF MEDIOCRITY

The principle of mediocrity is a rather general assumption (or speculation) that is often used in discussions concerning the nature and probability of extraterrestrial life. It assumes that things are pretty much the same all over—that is, it assumes that there is nothing special about Earth or humans' solar system. By invoking this hypothesis, scientists are suggesting that other parts of the universe are pretty much as they are here. This philosophical position allows them to then take the things they know about Earth, the chemical evolution of life that occurred here, and the facts that they are learning about other planetary bodies in this solar system and extrapolate this knowledge to develop concepts of what may be occurring on alien worlds around distant suns.

The simple premise of the principle of mediocrity is very often employed as the fundamental starting point for contemporary speculations about the cosmic prevalence of life. If Earth is indeed nothing *special*, then perhaps a million worlds in Milky Way galaxy (which is one of billions of galaxies) not only are suitable for the origin of life but also have witnessed its chemical evolution in their primeval oceans and are now (or at least were) habitats for a myriad of interesting living creatures. Some of these living systems may also have evolved to a level of intelligence such that the alien creatures are at this very moment gazing up into the heavens of their own world and wondering if they, too, are alone.

If, on the other hand, Earth and its intricately interwoven biosphere really are something special, then life—especially intelligent life that is capable of comprehending its own existence and contemplating its role in the cosmic scheme of things—may be a rare, very precious jewel in a vast, lifeless cosmos. Should Earth be unique—or Earth-like planets elsewhere be very rare—then the principle of mediocrity would be most inappropriate for use in estimating the probability that extraterrestrial life exists elsewhere in the universe.

Today, scientists cannot pass final judgment on the validity of the principle of mediocrity. They must, at an absolute minimum, wait until advanced robotic spacecraft and possibly human explorers have made more detailed investigations of the interesting other worlds in this solar system. Other worlds of particular interest to exobiologists include the planet Mars and certain moons of the giant outer planets Jupiter (especially its moon Europa) and Saturn (especially its moon Titan). Once scientists through their robot surrogates have explored these alien worlds in depth, they will have a much more accurate technical basis for suggesting that Earth is either "something special" or else "nothing special"—as the principle of mediocrity implies.

extrasolar planetary systems. To ascertain whether any of these extrasolar planets may be life sustaining, scientists will need to use advanced space-based systems—such as NASA's Kepler mission and the Space Interferometry Mission (SIM) in the next 10 years and the planned Terrestrial Planet Finder (TPF) and Life Finder missions another decade or so later. Instruments that are carried onboard spacecraft for these new missions will assist in the direct detection of terrestrial (that is, small and rocky) planetary companions to other stars, as well as investigate the composition of their atmospheres. (See chapters 3 and 7.) Liquid water is a basic requirement for life as scientists presently know and understand it. Specifically, during the next several decades, scientists will use advanced space-based instruments to detect key biomarkers that may strongly suggest whether planets revolving around other stars may indeed be life sustaining.

When viewed from a distance, planet Earth has known, readily identifiable surface biosignatures, or signs of life from changing vegetation patterns. In astronomy and exobiology, a *biosignature* is a spectral, photometric, or temporal signal whose origin specifically requires a biological agent. Earth also has several distinctive atmospheric biosignatures. Atmospheric biosignatures include the characteristic spectra of life-related molecular compounds such as oxygen (O_2), which is produced by photosynthetic plants and bacteria. (Chapter 3 provides more details about planetary biosignatures of importance in exobiology.)

This inspirational view of the "rising" Earth greeted the *Apollo 8* astronauts (Frank Borman, James A. Lovell, Jr., and William Anders) as they came from behind the Moon after performing the lunar-orbit insertion burn (December 1968). Each astronaut was awestruck by the stark contrast between barren and desolate lunar surface below them and the living, blue marble floating majestically in space in the distance. *(NASA)*

CONTINUOUSLY HABITABLE ZONE

According to astronomers and exobiologists, the continuously habitable zone (CHZ) is the region around a star in which one or several (Earth-like) planets can maintain conditions that are appropriate for the emergence and sustained existence of life. One important characteristic of a planet in the CHZ is that its environmental conditions support the retention of significant amounts of liquid water on the planet's surface. This potentially life-supporting region around a star is sometimes called the Goldilocks zone, and any Earth-like planet in the CHZ is often referred to as a Goldilocks planet. Displaying a fine sense of humor, scientists borrowed this unusual but appropriately descriptive nomenclature from the famous children's fairy tale, "Goldilocks and the Three Bears."

Within the context of contemporary astronomy and exobiology, an Earth-like planet is an extrasolar planet that is located in an ecosphere and has planetary environmental conditions that resemble the terrestrial biosphere—especially a suitable atmosphere, a temperature range that permits the retention of large quantities of liquid water on the planet's surface, and a sufficient quantity of radiant energy that strikes the planet's surface from the parent star. These suitable environmental conditions could permit the chemical evolution of carbon-based life, as scientists know it on Earth. The Earth-like planet should also have a mass somewhat greater than 0.4 Earth masses (to permit the production and retention of a breathable atmosphere) but less than about 2.4 Earth masses (to avoid excessive surface gravity conditions that could discourage the development of advanced life-forms—at least as known and understood here on Earth).

Perhaps an even more demanding question is: Does alien life (once started) develop to a level of intelligence? If exobiologists speculate that alien life does evolve to some level of intelligence, then they must also ask: Do intelligent alien life-forms acquire advanced technologies and learn to live with these vast powers over nature?

In sharp contrast to the vision of a universe full of emerging intelligent creatures, other scientists suggest that life itself is a very rare phenomenon and that human beings here on Earth are the only life-forms anywhere in the galaxy to have acquired a high-level of conscious intelligence and to have developed (potentially self-destructive) advanced technologies. Just think, for a moment, about the powerful implications of this latter conjecture. Are humans the best the universe has been able to produce in more than 14 billion years or so of cosmic evolution? If so, every human being is a very special living creature, not only on Earth but in the entire universe.

The preliminary scientific search on the Moon and on Mars for existing (or even extinct) extraterrestrial life has to date been unsuccessful and negative. However, as previously mentioned in chapter 1, the recent detailed study of a Martian meteorite by NASA scientists has renewed excitement and speculation about the possibility of microbial life on Mars—past or perhaps now precariously clinging to existence in some protected, subsurface biological niche.

Further encouraging the contemporary quest for microbial life beyond Earth is the recent discovery and recognition of *extremophiles* right here on humans' home planet. Terrestrial extremophiles are found in acid-rich hot springs, alkaline-rich soda lakes, and saturated salt beds. Additionally, examples of very hardy microbial life have been found in the Antarctic, living in rocks and at the bottoms of perennially ice-covered lakes. Life, including hardy microbial organisms, is also found in deep-sea hydrothermal vents at temperatures of up to 248°F (120°C). Bacteria have even been discovered in deep (0.6-mile [1-km] or deeper) subsurface ecosystems that derive their energy from basalt weathering. Some of these extremophiles can survive ultraviolet radiation or large doses of nuclear (ionizing) radiation, while others can tolerate extreme starvation, low nutrient levels, and low water activity. Remarkably, spore-forming bacteria have been revived from the stomachs of wasps, entombed in amber that was between 25 and 40 million years old. Clearly, life—at least as present on Earth—is diverse, tenacious, and adaptable to extreme environments.

Exobiologists postulate that under suitable environmental conditions, life, including intelligent life, should arise on planets around alien stars. This is an artist's rendering of an intelligent reptilelike creature. Some scientists suggest that on Earth, this type of very smart warm-blooded dinosaur might have evolved eventually had not a massive extinction event occurred some 65 million years ago, displacing the dinosaurs and allowing mammals, including *Homo sapiens,* to emerge. *(Department of Energy/Los Alamos National Laboratory)*

Results from the biology experiments onboard NASA's *Viking 1* and *2* lander spacecraft suggest that extant life is absent in surface environments on Mars. (Chapter 3 provides more details about NASA's amazing Viking Project.) However, life could be present in deep subsurface environments where liquid water may exist. Furthermore, although the present surface of Mars appears very inhospitable to life as we know it, recent space missions to the Red Planet have provided scientists with good evidence that the Martian surface environment was more Earth-like early in its history

EXTREMOPHILES

Extremophiles are hardy microorganisms that can exist under extreme physiochemical environmental conditions here on Earth, such as in frigid polar regions or boiling hot springs. Scientists generally characterize extremophiles according to the physical characteristics of the harsh environment in which they live and survive. For example, barophiles are microorganisms that thrive under high hydrostatic pressure conditions, as found typically in deep marine environments; while thermophiles are microorganisms that can live and grow in high temperature environments. Invoking the principle of mediocrity, some exobiologists have speculated that similar hardy (extraterrestrial) microorganisms might exist elsewhere in this solar system, perhaps within subsurface biological niches on Mars or in the suspected liquid water ocean beneath the frozen surface of Europa.

This artist's rendering depicts extremely hostile and diverse environments that might be found on extrasolar planets or companion moons. Biologists have discovered very hardy microorganisms, called extremophiles, surviving and even thriving in extreme environments here on Earth. So exobiologists suggest that similar hardy microorganisms might arise and survive in the harsh environments of extrasolar planetary systems around alien stars. *(NASA; artist, Pat Rawlings)*

(some 3.5 to 4.0 billion years ago), with a warmer climate and liquid water at or near the surface. Scientists know that life originated very quickly on the early Earth (perhaps within a few hundred million years), and so it seems quite reasonable to assume that life could have emerged on Mars during a similar early window of opportunity when liquid water was present at the surface.

Of course, the final verdict concerning life on Mars (past or present) will not be resolved properly until more detailed investigations of the Red Planet takes place. Perhaps later this century, a terrestrial explorer (robot or human) will stumble upon a remote exobiological niche in some deep Martian canyon, or possibly a team of astronaut-miners, searching for certain ores on Mars, will uncover the fossilized remains of a tiny ancient creature that roamed the surface of the Red Planet in more hospitable environmental eras. Speculation, yes—but not without reason. (Chapter 4 provides more discussion about the search for life on Mars.)

The giant outer planets and their constellations of intriguing moons also present some tantalizing possibilities for extraterrestrial life. Which exobiologist cannot become excited about the possible existence of an ocean of liquid water beneath the frozen surface of the Jovian moon Europa and the (remote) chance that this alien-world ocean might contain communities of extraterrestrial life-forms clustered around hydrothermal vents? Furthermore, data from NASA's *Galileo* spacecraft suggest that two other Jovian moons, Callisto and Ganymede, may also have liquid water deposits beneath their icy surfaces. (See chapter 5.)

All scientists can say at this point with any degree of certainty is that their overall understanding of the cosmic prevalence of life will be significantly influenced by the exobiological discoveries (pro and con) that will occur in the next few decades both on planetary bodies in the solar system and concerning extrasolar planets beyond.

Recent discoveries show that comets could represent a unique repository of information about chemical evolution and organic synthesis at the very outset of the solar system. After reviewing comet Halley encounter data, space scientists suggested that comets have remained unchanged since the formation of the solar system. Exobiologists now have evidence that the organic molecules considered to be the molecular precursors to those essential for life are prevalent in comets. These discoveries have provided further support for the hypothesis that the chemical evolution of life has occurred and is now occurring widely throughout the Milky Way galaxy. Some scientists even suggest that comets have played a significant role in the chemical evolution of life on Earth. They hypothesize that significant quantities of important life-precursor molecules could have been deposited in an ancient terrestrial atmosphere by cometary collisions.

Meteoroids are solid chunks of extraterrestrial matter. As such, they represent another source of interesting information about the occurrence of prebiotic chemistry beyond the Earth. In 1969, meteorite analysis provided the first convincing proof of the existence of extraterrestrial amino acids. Amino acids are a group of molecules necessary for life. Since that time, a large amount of information has been gathered that shows that many more of the molecules that are considered necessary for life are also present in meteorites. As a result of this line of investigation, it now seems clear to exobiologists that the chemistry of life may not and should

Evidence is now mounting that early in its history, the planet Mars possessed flowing water on its surface—a condition that could have led to the development of life. This is an artist's rendering of what an ancient Mars might have looked like with liquid water on its surface. Detailed exploration of Mars in the first two decades of this century will help confirm this hypothesis and possibly uncover indisputable signs of life—extinct or perhaps even extant in some sheltered subsurface biological niche. (NASA/JPL)

not be unique to Earth. Future work in this area will greatly help scientists develop a better understanding of the conditions and processes that existed during the formation of the solar system. These studies should also provide clues concerning the relations between the origin of the solar system and the origin of life.

The basic question—"Is life—especially intelligent life—unique to the Earth?"—lies at the very core of our concept of self and where humans fit in the cosmic scheme of things. If life is extremely rare, then all members of the human race have a truly serious collective obligation to the entire (as yet "unborn") universe. As an intelligent species, we must preserve carefully the precious biological heritage that has taken more than four billion years to evolve on this tiny planet. If, on the other hand, life (including intelligent life) is abundant throughout the galaxy, then human beings should eagerly seek to learn of its existence and ultimately should become part of a galactic family of conscious, intelligent creatures. Fermi's famous paradoxical question "Where are they?" takes on special significance this century. (See chapter 8.)

✧ Panspermia

Panspermia is the general hypothesis that microorganisms, spores, or bacteria attached to tiny particles of matter have diffused through space, eventually encountering a suitable planet and initiating the rise of life there. The word itself means "all-seeding."

In the 19th century, the Scottish scientist Lord Kelvin (Baron William Thomson [1824–1907]) suggested that life could have arrived here on Earth from outer space, perhaps carried inside meteorites. Then, in 1908, with the publication of his book *Worlds in the Making,* the Swedish chemist and the Nobel laureate Svante August Arrhenius (1859–1927) put forward the idea that is now generally regarded as the *panspermia hypothesis.* Arrhenius said that life really did not start here on Earth but rather was "seeded" by means of extraterrestrial spores (seedlike germs), bacteria, or microorganisms. According to his hypothesis, these microorganisms, spores, or bacteria originated elsewhere in the Milky Way galaxy (possibly on a planet in another star system where conditions were more favorable for the chemical evolution of life) and then wandered through space attached to tiny bits of cosmic matter rather than moved under the influence of stellar radiation pressure.

The greatest difficulty most scientists have today with Arrhenius's original panspermia concept is simply the question of how these "life-seeds" can wander through interstellar space for up to several billion years, receive extremely severe radiation doses from cosmic rays, and still be "vital" when they eventually encounter a solar system that contains suitable planets. Even on a solar-system scale, the survival of such micro-organisms, spores, or bacteria would be difficult. For example, life seeds wandering from the vicinity of Earth to Mars would be exposed to both ultraviolet radiation from the Sun and ionizing radiation in the form of solar-flare particles and cosmic rays. The interplanetary migration of spores might take several hundred thousand years in the airless, hostile environmental conditions of outer space.

Nobel laureate Francis Crick (1916–2004) and Leslie Orgel (1927–   ) attempted to resolve this difficulty by proposing the *directed-panspermia hypothesis.* Feeling that the overall concept of panspermia was too interest-ing to abandon entirely, in the early 1970s, they suggested that an ancient, intelligent alien race could have constructed suitable interstellar robot spacecraft; loaded these vehicles with an appropriate cargo of microor-ganisms, spores, or bacteria; and then proceeded to "seed the galaxy" with life, or at least the precursors of life. This life-seed cargo would have been protected during the long interstellar journey and then released into suitable planetary atmospheres or oceans when favorable planets were encountered by the robot starships.

Why would an extraterrestrial civilization undertake this type of proj-ect? It might first have tried to communicate with other races across the interstellar void. Then, when this failed, the alien civilization could have convinced itself that it was *alone.* At this point in its civilization, driven by some form of "missionary zeal" to "green" (or perhaps "blue") the Milky Way galaxy with life as that intelligent alien species knew it, alien scientists might have initiated a sophisticated directed-panspermia program. Smart robot spacecraft containing well-protected spores, microorganisms, or bacteria were launched into the interstellar void to seek new "life sites" in neighboring star systems. This effort might have been part of an advanced-technology demonstration program—a form of planetary engineering on an interstellar scale. These life-seeding robot spacecraft may also have been the precursors of an ambitious colonization wave that never came—or perhaps is just now getting under way.

In their directed-panspermia discussions, Crick and Orgel identified what they called the theorem of detailed cosmic reversibility. This theorem suggests that if humans can now contaminate other worlds in our solar system with microorganisms that are hitchhiking on terrestrial spacecraft, then it is also reasonable to assume that an advanced, intelligent extrater-restrial civilization could have used its robot spacecraft to contaminate or seed other worlds (including Earth) with spores, microorganisms, or bacteria sometime in the very distant past.

Other scientists have suggested that life on Earth might have evolved as a result of microorganisms that were left here inadvertently by ancient astronauts themselves. It is most amusing to speculate that humans may be here today because ancient space travelers were "litterbugs," scattering their garbage on a then-lifeless planet. This line of speculation is sometimes called the extraterrestrial garbage theory of the origin of terrestrial life.

Sir Fred Hoyle (1915–2001) and Nalin Chandra (N. C.) Wickramasinghe (1939–   ) have also explored the issue of directed panspermia and the origin of life on Earth. In several publications, they argued convincingly that the biological composition of living things on Earth has been and will continue to be influenced radically by the arrival of "pristine genes" from space. They further suggested that the arrival of these cosmic microorganisms, and the resultant complexity of terrestrial life, is not a random process but is one carried out under the influence of a greater cosmic intelligence.

This brings up another interesting point. As scientists and engineers here on Earth develop the technology necessary to send smart machines and human explorers to other worlds in this solar system (and eventually beyond to other star systems), should future generations of humans initiate a program of directed panspermia? If our descendents became convinced that human beings might really be alone in the galaxy, then strong intellectual and biological imperatives might urge them to start the process of "greening the galaxy" or to seed life where there is now none. (See chapter 10.)

Perhaps late in this century, robot interstellar explorers will be sent from humans' solar system, not only to search for extraterrestrial life but also to plant life on potentially suitable extrasolar planets when no life is found. This may be one of humans' higher cosmic callings—to be the first intelligent species to develop to a level of technology that permits the expansion of life itself within the galaxy. Of course, the human race's directed-panspermia effort might only be the next link in a cosmic chain of events that was started eons ago by a long-since extinct alien civilization. Millions of years from now, on an Earth-like planet around a distant sunlike star, other intelligent beings will start wondering whether life on their world started spontaneously or was seeded there by an ancient civilization (in this case, *terrestrial*) that has long since disappeared from view in the galaxy. While the panspermia or directed-panspermia hypotheses do not address how life originally started somewhere in the galaxy, they certainly provide some intriguing concepts regarding how, once started, life might "get around."

✧ We Are Made of Stardust

Throughout most of human history, people considered themselves and the planet they lived on as being apart from the rest of the universe. After all, the heavens were clearly unreachable and therefore had to remain the abode of the deities found in the numerous mythologies that enriched ancient civilizations. It is only with the rise of modern astronomy, exobiology, and space technology that scientists have been able to investigate properly the chemical evolution of the universe. And the results are nothing short of amazing.

While songwriters and poets often suggest that a loved one is made of stardust, modern scientists have shown us that this is *not* just a fanciful artistic expression; it is quite literally true. All of us are made of stardust! Thanks to a variety of astrophysical phenomena, including ancient stellar explosions that took place long before the solar system formed, the chemical elements enriching our world and supporting life came from the stars. This section provides a brief introduction to the cosmic connection of the chemical elements.

The chemical elements, such as carbon (C), oxygen (O), and calcium (Ca), are all around us and are part of us. Furthermore, the composition of planet Earth and the chemical processes that govern life within our planet's biosphere are rooted in these chemical elements. To acknowledge the relationship between the chemical elements and life, scientists have given a special name to the group of chemical elements that they consider to be essential for all living systems—whether here on Earth, possibly elsewhere in the solar system, or perhaps on habitable planets around other stars. As previously mentioned, scientists refer to this special group of life-sustaining chemical elements as the biogenic elements.

Biologists focus their studies on life as it occurs on Earth in its numerous, greatly varied, and interesting forms. Exobiologists extend basic concepts about carbon-based life here on Earth to create their scientifically based speculations about the possible characteristics of life beyond the terrestrial biosphere. When considering the biogenic elements, scientists usually place primary emphasis on the elements hydrogen (H), carbon, nitrogen (N), oxygen, sulfur (S), and phosphorous (P). The chemical compounds of major interest are those normally associated with water (H_2O) and with other organic chemicals in which carbon bonds with itself or with other biogenic elements. There are also several "life-essential" inorganic chemical elements, including iron (Fe), magnesium (Mg), calcium, sodium (Na), potassium (K), and chlorine (Cl).

All the natural chemical elements found here on Earth and elsewhere in the universe have their ultimate origins in cosmic events. Since different elements come from different events, the elements that make up life itself reflect a variety of astrophysical phenomena that have taken place in the universe. For example, the hydrogen found both in water and hydrocarbon molecules formed just a few moments after the big bang event that started the universe. Carbon, the element that is considered to be the basis for all terrestrial life, formed in small stars. Elements such as calcium and iron formed in the interiors of large stars. Heavier elements with atomic numbers beyond iron, such as silver (Ag) and gold (Au), formed in the tremendous explosive releases of supernovae. Certain light elements, such as lithium (Li), beryllium (Be), and boron (B), resulted from energetic cosmic-ray interactions with other atoms, including the hydrogen or helium nuclei that are found in interstellar space.

Following the big bang explosion, the early universe contained the primordial mixture of energy and matter that evolved into all the forms of energy and matter that scientists observe in the universe today. For example, about 100 seconds after the big bang, the temperature of this expanding mixture of matter and energy fell to approximately 1 billion kelvins—"cool" enough so that neutrons and protons began to stick to each other during certain collisions and form light nuclei, such as deuterium and lithium. When the universe was about three minutes old, 95 percent of the nuclei were hydrogen, 5 percent were helium, and there were only trace amounts of lithium. At the time, the nuclei of these three light elements were the only ones that existed.

As the universe continued to expand and cool, the early atoms (mostly hydrogen and a small amount of helium) began to form through the capture of electrons and then gather through gravitational attraction into very large clouds of gas. For millions of years these giant gas clouds were the only matter in the universe because neither stars nor planets had yet formed. Then, about 200 million years after the big bang, the first stars began to shine, and the creation of important new chemical elements started in their thermonuclear furnaces.

Stars form when giant clouds of mostly hydrogen gas, perhaps light-years across, begin to contract under the attractive force of their own gravity. During millions of years, various clumps of hydrogen gas would eventually collect into a giant ball of gas that was hundreds of thousands of times more massive than Earth. As the giant gas ball continued to contract under its own gravitational influence, an enormous pressure arose in its interior. Consistent with the laws of physics, the increase in pressure at the center of this "protostar" was accompanied by an increase in temperature. Then, when the center reached a minimum temperature of about 15 million kelvins, the hydrogen nuclei in the center of the contracting gas ball moved fast enough so that when they collided, these light (low-mass) atomic nuclei would undergo fusion. This is the very special moment when a new star is born.

The process of nuclear fusion releases a great amount of energy at the center of the star. Once thermonuclear burning begins in a star's core, the internal energy release counteracts the continued contraction of stellar mass by gravity. The ball of gas becomes stable—as the inward pull of gravity exactly balances the outward radiant pressure from thermonuclear fusion reactions in the core. Ultimately, the energy released in fusion flows upward to the star's outer surface, and the new star "shines." It is this continuous, radiant outflow of energy from a parent star that provides the energy necessary to sustain life on any habitable planets that may orbit the star.

Stars come in a variety of sizes, ranging from about one-tenth to 60 (or more) times the mass of our parent star, the Sun. It was not until the mid-1930s that astrophysicists began to recognize how the process of nuclear fusion takes place in the interiors of all normal stars and fuels their enormous radiant energy outputs. Scientists use the term *nucleosynthesis* to describe the complex process of how different size stars create different elements through nuclear fusion reactions.

Astrophysicists and astronomers consider stars that are less than about five times the mass of the Sun to be medium- and small-size stars. The production of elements in stars within this mass range is similar. Small- and medium-size stars also share a similar fate at the end of life. At birth, small stars begin their stellar life by fusing hydrogen into helium in their cores. This process generally continues for billions of years, until there is no longer enough hydrogen in a particular stellar core to fuse into helium. Once hydrogen burning stops, so does the release of the thermonuclear energy that produced the radiant pressure, which counteracted the relentless inward attraction of gravity. At this point in its life, a small star begins to collapse inward. Gravitational contraction causes an increase in temperature and pressure. As a consequence, any hydrogen remaining in the star's middle layers soon becomes hot enough to undergo thermonuclear fusion into helium in a "shell" around the dying star's core. The release of fusion energy in this shell enlarges the star's outer layers, causing the star to expand far beyond its previous dimensions. This expansion process cools the outer layers of the star, transforming them from brilliant white hot or bright yellow in color to a shade of dull glowing red. Quite understandably, astronomers call a star at this point in its life cycle a red giant.

Gravitational attraction continues to make the small star collapse, until the pressure in its core reaches a temperature of about 100 million kelvins. This very high temperature is sufficient to allow the thermonuclear fusion of helium into carbon. The fusion of helium into carbon now releases enough energy to prevent further gravitational collapse—at least until the helium runs out. This stepwise process continues until oxygen is fused. When there is no more material to fuse at the progressively increasing high temperature conditions within the collapsing core, gravity again exerts it relentless attractive influence on matter. This time, however, the heat released during gravitational collapse causes the outer layers of the small star to blow off, creating an expanding symmetrical cloud of material that astronomers call a planetary nebula. The expanding cloud may contain as much as 10 percent of the small- or medium-size star's mass. The explosive blow-off process is very important because it disperses into space the elements created in the small star's core by nucleosynthesis.

The final collapse that causes the small star to eject a planetary nebula also liberates thermal energy. But this time, the energy release is not enough to fuse other elements, so the remaining core material continues to collapse until all the atoms are crushed together and only the repulsive force between the electrons counteracts gravity's relentless pull. Astronomers refer to this type of condensed matter as a degenerate star and give the final compact object a special name—the white dwarf star. The white dwarf star represents the final phase in the evolution of most low mass stars, including the Sun.

If the white dwarf star is a member of a binary star system, its intense gravity might pull some gas away from the outer regions of the companion (normal) star. When this happens, the intense gravity of the white dwarf causes the inflowing new gas to reach rapidly very high temperatures, and a sudden explosion occurs. Astronomers call this event a nova. The nova explosion can make a white dwarf appear up to 10,000 times brighter for a short period of time. Thermonuclear fusion reactions that take place during the nova explosion also create new elements, such as carbon, oxygen, nitrogen, and neon. These elements are then dispersed into space.

In some very rare cases, a white dwarf might undergo a gigantic explosion that astrophysicists call a Type Ia supernova. This happens when a white dwarf is part of a binary star system and pulls too much matter from its stellar companion. Suddenly, the compact star can no longer support the additional mass, and even the repulsive pressure of electrons in crushed atoms can no longer prevent further gravitational collapse. This new wave of gravitational collapse heats the helium and carbon nuclei in a white dwarf and causes them to fuse into nickel, cobalt, and iron. However, the thermonuclear burning now occurs so fast that the white dwarf completely explodes. During this rare occurrence, nothing is left behind. All the elements created by nucleosynthesis during the lifetime of the small star now scatter into space as a result of this spectacular supernova detonation.

Large stars have more than five times the mass of the Sun. These stars begin their lives in pretty much the same way as small stars—by fusing hydrogen into helium. However, because of their size, large stars burn faster and hotter, generally fusing all the hydrogen in their cores into helium in less than one billion years. Once the hydrogen in the large star's core is fused into helium, it becomes a red supergiant—a stellar object similar to the red giant star previously mentioned, only much larger.

However, unlike a red giant, the enormous red supergiant star has enough mass to produce much higher core temperatures as a result of gravitational contraction. A red supergiant fuses helium into carbon, carbon and helium into oxygen, and even two carbon nuclei into magnesium. Thus, through a combination of intricate nucleosynthesis reactions, the supergiant star forms progressively heavier elements up to and including the element iron. Astrophysicists suggest that the red supergiant has an onionlike structure—with different elements being fused at different temperatures in layers around the core. The process of convection brings these elements from the star's interior to near its surface, where strong stellar winds then disperse them into space.

Thermonuclear fusion continues in a red supergiant star until the element iron is formed. Iron is the most stable of all the elements. Elements lighter than (below) iron on the periodic table generally emit energy when joined or fused in thermonuclear reactions, while elements heavier than (above) iron on the periodic table emit energy only when their nuclei split or fission. So from where did the elements that are more massive than iron come? Astrophysicists postulate that neutron capture is one way in which the more massive elements form. Neutron capture occurs when a free neutron (one outside its parent atomic nucleus) collides with another atomic nucleus and "sticks." This capture process changes the nature of the compound nucleus, which is often radioactive and undergoes decay, thereby creating a different element with a new atomic number.

While neutron capture can take place in the interior of a star, it is during a supernova explosion that many of the heavier elements, such as iodine, xenon, gold, and the majority of the naturally occurring radioactive elements, are formed by numerous, rapid neutron capture reactions.

What happens when a large (greater than five solar masses) star goes supernova? The red supergiant eventually produces the element iron in its intensely hot core. However, because of nuclear stability phenomena, the element iron is the last chemical element formed in nucleosynthesis. When fusion begins to fill the core of a red supergiant star with iron, thermonuclear energy release in the large star's interior decreases. Because of this decline, the star no longer has the internal radiant pressure to resist the attractive force of gravity. And so the red supergiant begins to collapse. Suddenly, this gravitational collapse causes the core temperature to rise to more than 100 billion kelvins, smashing the electrons and protons in each iron atom together to form neutrons. The force of gravity now draws this massive collection of neutrons incredibly close together. For about a second, the neutrons fall very fast toward the center of the star. Then they smash into each other and suddenly stop. This sudden stop causes the neutrons to recoil violently, and an explosive shock wave travels outward from the highly compressed core. As this shock wave travels from the core, it heats and accelerates the outer layers of material of the red supergiant star. The traveling shock wave causes the majority of the large star's mass to be blown off into space. Astrophysicists call this enormous explosion a Type II supernova.

A supernova will often release (for a brief moment) enough energy to outshine an entire galaxy. Since supernovae explosions scatter elements made within red supergiant stars far out into space, they are one of the most important ways the chemical elements disperse in the universe. Just before the outer material is driven off into space, the tremendous force of the supernova explosion provides nuclear conditions that support rapid capture of neutrons. Rapid neutron capture reactions transform elements in the outer layers of the red supergiant star into radioactive isotopes that decay into elements heavier than iron.

This section could only provide a very brief glimpse of our cosmic connection to the chemical elements. But the next time you look up at the stars on a clear night, just remember that you, all other persons, and everything in our beautiful home planet is made of stardust.

✧ Is Earth the Galaxy's Most Unusual and Lucky Planet?

Earth is the third planet from the Sun and the fifth-largest in the solar system. Our home planet circles its parent star at an average distance of about 93 million miles (149.6 million km). Earth is one of four so-called terrestrial planets found in the inner portion of the solar system. In addition to Earth itself, the terrestrial (or inner) planets are Mercury, Venus, and Mars. These planets are similar in their physical properties and characteristics to Earth—that is, they are small, relatively high-density bodies that are composed of metals and silicates with shallow (or negligible) atmospheres as compared to the giant, gaseous outer planets (Jupiter, Saturn, Uranus, and Neptune). What makes Earth so unique (or perhaps lucky) is that it is the only planetary body in the solar system currently known to support life. As discussed later in the book, Mars and the Jovian moon Europa are suspected of possibly being life-bearing worlds, but at present, only our home planet is known to host a treasure house of biological activity.

The name *Earth* comes from the Indo-European language base *er,* which produced the Germanic noun *ertho,* and ultimately the German word *erde,* the Dutch *aarde,* the Scandinavian *jord,* and the English *earth.* Related word forms include the Greek *eraze,* meaning on the "ground," and the Welsh *erw,* meaning "a piece of land." In Greek mythology, the goddess

Life on Earth comes in a wide variety of forms, shapes, and sizes. For example, this stingray is just one of the many kinds of marine life that inhabit the waters off the Florida Keys. *(NASA)*

of Earth was called *Gaia*, while in Roman mythology the Earth goddess was *Tellus* (meaning "fertile soil"). The expression *Mother Earth* comes from the Latin expression *terra mater*. Scientists and writers frequently use the word *terrestrial* to describe creatures and things related to or from planet Earth and the word *extraterrestrial* to describe creatures and things beyond or away from planet Earth. Astronomers also refer to our planet as *Terra* or *Sol III*, which means the third satellite out from the Sun.

From space, Earth is characterized by its blue waters and white clouds, which cover a major portion of it. An ocean of air, consisting of 78 percent nitrogen and 21 percent oxygen, surrounds humans' home planet; the remainder is argon, neon, and other gases. The standard atmospheric pressure at sea level is 14.7 pounds per square inch (101,325 Pa). Surface temperatures range from a maximum of about 140°F (60°C) in desert regions along the equator to a minimum of -130°F (-90°C) in the frigid, polar regions. In between, however, surface temperatures are generally much more benign. The acceleration due to gravity at sea level on Earth is 32.2 feet per second-squared (9.8 m/s²). Some exobiologists suggest that a surface gravity value much larger or smaller than this value would not favor the emergence of complex, possible intelligent life.

Earth's rapid spin and molten nickel-iron core give rise to an extensive magnetic field. This magnetic field, together with the atmosphere, shields people and other living creatures from nearly all of the harmful charged particles and ultraviolet radiation coming from the Sun and other cosmic sources. Furthermore, most meteors burn up in Earth's protective atmosphere before they can strike the surface. Earth's nearest celestial neighbor, the Moon, is its only natural satellite.

While life on Earth is made possible by the Sun, terrestrial life is also regulated by the periodic motions of the Moon. The ocean tides rise and fall because of the gravitational tug-of-war between Earth and the Moon. Throughout history, the Moon has had a significant influence on human culture, art, and literature. For example, the months of the calendar year originated from the regular motions of the Moon around Earth. Even in the space age, the Moon has proved to be a major technical stimulus. The Moon was just far enough away to represent a real technical challenge to reach it; yet this alien world was close enough to allow humans to be successful on the first concentrated effort.

The most recent lunar origin theory suggests a cataclysmic birth of the Moon. Scientists who support this theory speculate that near the end of Earth's accretion from the primordial solar nebula materials (that is, after its core was formed, but while Earth was still in a molten state), a Mars-size celestial object (called an impactor) hit Earth at an oblique angle. This ancient explosive collision sent vaporized-impactor and molten-Earth material into Earth orbit, and the Moon then formed from these materials. Was the formation of Earth's large natural satellite a lucky, very rare random cosmic event, or does this type of cosmic coincidence happen frequently to Earth-like planets around distant stars? The scientific answer to this interesting question may shed additional light on the overall issue of the life, especially intelligent life, in other star systems.

✧ Gaia Hypothesis

The Gaia hypothesis was first suggested in 1969 by the British biologist James Lovelock (1919–)—with the assistance of the biologist Lynn Margulis (1938–). This interesting hypothesis states that Earth's biosphere has an important modulating effect on the terrestrial atmosphere. Because of the chemical complexity observed in the lower atmosphere, Lovelock postulated that life-forms within the terrestrial biosphere actually help control the chemical composition of the Earth's atmosphere, thereby ensuring the continuation of conditions suitable for life. Gas-exchanging microorganisms, for example, are thought to play a key role in this continuous process of environmental regulation. Without these "cooperative" interactions in which some organisms generate certain gases and carbon compounds that are subsequently removed and used by other organisms, planet Earth might also possess an excessively hot or cold planetary surface that is devoid of liquid water and is surrounded by an inanimate, carbon dioxide–rich atmosphere.

Gaia (also spelled *Gaea*) was the goddess of Earth in ancient Greek mythology. Lovelock used her name to represent the terrestrial biosphere—namely, the system of life on Earth, including living organisms and their required liquids, gases, and solids. Thus, the Gaia hypothesis metaphorically implies that "Gaia" (Earth's biosphere) will struggle to maintain the atmospheric conditions that are suitable for the survival of terrestrial life.

If scientists use the Gaia hypothesis in their search for extraterrestrial life, they should look for extrasolar planets that exhibit variability in atmospheric composition. Extending this hypothesis beyond the terrestrial biosphere, a planet will either be living or else it will not. The absence of chemical interactions in the lower atmosphere of an alien world could be taken as an indication of the absence of living organisms.

Although this interesting hypothesis is currently more speculation than hard, scientifically verifiable fact, it is still quite useful in developing a sense of appreciation for the complex chemical interactions that have helped to sustain life in the Earth's biosphere. These interactions among microorganisms, higher-level animals, and their mutually shared atmosphere might also have to be carefully considered in the successful development of effective closed life-support systems for use on permanent space stations, lunar bases, and planetary settlements.

✧ Extraterrestrial Catastrophe Theory

For millions of years, giant, thundering reptiles roamed the lands, dominated the skies, and swam in the oceans of a prehistoric Earth. Dinosaurs reigned supreme. Then, quite suddenly, some 65 million years ago, they vanished. What happened to these giant creatures and to thousands of other ancient animal species?

From archaeological and geological records, scientists know that some tremendous catastrophe occurred about 65 million years ago on this planet. It affected life more extensively than any war, famine, or plague in human history, for in that cataclysm, about 70 percent of all species then living on Earth—including, of course, the dinosaurs—disappeared within a very short period. This mass extinction is also referred to Cretaceous-Tertiary Mass Extinction event, or simply the K-T event.

In 1980, the scientists Luis W. Alvarez (1911–88) and his son, Walter Alvarez (1940–   ), along with their colleagues at the University of California at Berkeley, discovered that a pronounced increase in the amount of the element iridium in Earth's surface had occurred at precisely the time of the disappearance of the dinosaurs. First seen in a peculiar sedimentary-clay area found near Gubbio, Italy, the same iridium enhancement was soon discovered in other places around the world in the thin sedimentary layer that was laid down at the time of the mass extinction. Since iridium is quite rare in Earth's crust and is more abundant in the rest of the solar system, the Alvarez team postulated that a large asteroid (about 6 miles [10 km] or more in diameter) had struck the ancient Earth. This cosmic collision would have promoted an environmental catastrophe throughout the planet. The scientists reasoned that such an asteroid would largely vaporize while passing through the Earth's atmosphere, spreading a dense cloud of dust particles, including quantities of extraterrestrial iridium atoms, uniformly around the globe.

Stimulated by the Alvarez team's postulation, many subsequent geologic investigations have observed a global-level of enhanced iridium in this thin layer (about 0.4-inch [1-cm] thick) of the Earth's crust (lithosphere) that lies between the final geologic formations of the Cretaceous period (which are dinosaur-fossil rich) and the formations of the early Tertiary period (whose rocks are notably lacking in dinosaur fossils). The Alvarez hypothesis further speculated that following this asteroid impact, a dense cloud of dust covered Earth for many years, obscuring the Sun, blocking photosynthesis, and destroying the very food chains upon which many ancient life-forms depended.

Despite the numerous geophysical observations of enhanced iridium levels that reinforced the Alvarez team's impact hypothesis, many geologists and paleontologists still preferred other explanations concerning the mass extinction that occurred about 65 million years ago. To them, the impact theory of mass extinction was still a bit untidy. Where was the impact crater? This important question was answered in the early 1990s when a 112-mile- (180-km-) diameter ring structure called Chicxulub was identified from geophysical data collected in the Yucatán region of Mexico. The Chicxulub crater has been age-dated at 65 million years. Further studies have also helped confirm its impact origin. The impact of a 6-mile- (10-km-) diameter asteroid would have created this very large crater, as well as causing enormous tidal waves. Scientists have found evidence of tidal waves occurring about 65 million years ago around the Gulf of Mexico region.

Of course, there are still many other scientific opinions as to why the dinosaurs vanished. A popular one is that there was a gradual but relentless change in the Earth's climate to which these giant reptiles and many other prehistoric animals simply could not adapt. So, one can never absolutely prove that an asteroid impact "killed the dinosaurs." Many species of dinosaurs (and smaller flora and fauna) had, in fact, become extinct during the millions of years preceding the K-T event. However, the impact of a 6-mile- (10-km-) across asteroid would most certainly have been an immense insult to life on Earth. Locally, there would have been intense shock-wave heating and fires, tremendous earthquakes, hurricane winds, and hundreds of billions of tons of debris thrown everywhere. This debris would have created months of darkness and cooler temperatures on a global scale. There would also have been concentrated nitric acid rains worldwide. Sulfuric acid aerosols may have cooled Earth for years after the impact. Life certainly would not have been easy for those species that did survive. Fortunately, such large, extinction-level events (ELEs) are thought to occur only about once every hundred million years. It is also interesting to observe, however, that as long as those enormous reptiles roamed and dominated the Earth, mammals, including humans themselves, would have had little chance of evolving. So if the large asteroid impact hypoth-

An artist's rendering of a killer impactor striking a coastal region on Earth. All life near the impact point is instantly destroyed from the effects of high temperatures and pressures. Giant tidal waves would carry destruction well beyond the impact zone—up and down the doomed coastline, as well as far inland. Finally, huge quantities of dust from this impact would be hurled high into the upper atmosphere, blocking life-sustaining sunlight and creating a planetwide "nuclear winter" that causes the eventual extinction of most species. *(NASA; artist, Don Davis [artwork created May 9, 1992])*

esis is correct, the ancient catastrophic event was clearly a case of bad luck for the dinosaurs. Yet, this K–T event was also a case of exceptionally good fortune for the eventual emergence of human beings—the intelligent creatures who would dominate the planet some 65 million years later and start to travel beyond Earth into the solar system.

The possibility that an asteroid or comet will strike Earth in the future is quite real. Just look at an image of Mars or the Moon, and ask how those large impact craters were formed. A comet, called Shoemaker-Levy 9, hit Jupiter in 1994. Fortunately, the probability of a really *large* asteroid or comet striking the Earth is quite low. For example, space scientists estimate that Earth will experience (on average) one collision with an Earth-crossing asteroid (ECA) of 0.6-mile- (1-km-) diameter size or greater every 300,000 years.

Yet, on May 22, 1989, a small ECA, called 1989FC, passed within 428,840 miles (690,000 km) of humans' home planet. This cosmic "near-miss" occurred with just 0.0046 astronomical unit to spare—a distance less than twice the distance to the Moon. Cosmic impact specialists have estimated that if this small asteroid, presumed to be about 660 feet (200 m) to 1,310 feet (400 m) in diameter, had experienced a straight-in collision with Earth at a relative velocity of some 10 miles per second (16 km/s), it would have impacted with an explosive force of between 400 to 2,000 megatons (MT). A megaton is the energy of an explosion that is the equivalent to one million tons of the chemical high explosive trinitrotoluene (TNT). If this small asteroid had hit a terrestrial landmass, it would have formed a crater some 2.5-miles (4-km) to 4.4-miles (7-km) across and produced a great deal of regional-scale (but probably not global-scale) destruction.

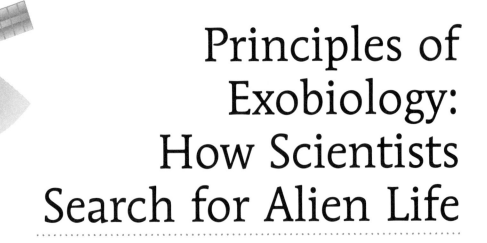

Principles of Exobiology: How Scientists Search for Alien Life

In its most general definition, *exobiology* is the study of the living universe. The term *astrobiology* is sometimes used for exobiology. Contemporary exobiologists are concerned with observing and exploring outer space to answer the following intriguing questions (among many others): Where did life come from? How did it evolve? Where is the evolution of life leading? Is the presence of life on Earth unique, or is it a common phenomenon that occurs whenever suitable stars evolve with companion planetary systems?

Looking within this solar system, exobiologists inquire: Did life evolve on Mars? If so, is it now extinct, or does it still exist there—perhaps clinging to survival in some remote, subsurface location on the Red Planet? Does Europa, one of the intriguing ice-covered moons of Jupiter, now harbor life within its anticipated subsurface liquid-water ocean? What about the possibility of life in the subsurface liquid-water oceans now suspected to exist on Callisto and Ganymede—two other planet-sized icy moons of Jupiter? Exobiologists are also trying to understand the significance of the complex organic molecules that appear to be forming continuously in the nitrogen-rich atmosphere of Saturn's moon Titan.

Beyond Earth, yet within this solar system, scientists are using a variety of sophisticated robot spacecraft to continue their search for life. With each sophisticated space mission during this century, scientists seek new evidence about the possible origin of life on Mars or Europa. Data from Titan, as well as from solar system small bodies such as comets and asteroids, are also providing valuable clues about conditions in the early solar system that promoted the development of life here on Earth—and possibly elsewhere. The more scientists learn about these diverse worlds, the more improved becomes their overall ability to search for other, possibly life-bearing, worlds around distant stars.

Exobiology is more rigorously defined as the multidisciplinary field that involves the study of extraterrestrial environments for living organisms, the recognition of evidence of the possible existence of life in these environments, and the study of any extraterrestrial life-forms that may be encountered. As mentioned in chapter 2, biophysicists, biochemists, and exobiologists usually define a *living organism* as a physical system that exhibits the following general characteristics: It has structure (that is, it contains information), can replicate itself, and experiences few (random) changes in its information package, which supports a Darwinian evolution (i.e., survival of the fittest).

Observational exobiology involves the detailed investigation of other interesting celestial bodies in this solar system, as well as the search for extrasolar planets and the study of the organic chemistry of interstellar molecular clouds. *Exopaleontology* involves the search for the fossils or biomarkers of extinct alien life-forms in returned soil and rock samples, unusual meteorites, or in-situ (on other worlds) by robot and/or human explorers. *Experimental exobiology* includes the evaluation of the viability of terrestrial microorganisms in space and the adaptation of living organisms to different (planetary) environments.

The challenges of exobiology can be approached from several different directions. First, pristine material samples from interesting alien worlds in our solar system can be obtained for study on Earth—as was accomplished during the Apollo Project lunar expeditions (1969–72)—or such samples can be studied on the spot (in-situ) by robot explorers—as was accomplished by the robot Viking landers (1976). The returned lunar rock and soil samples have not revealed any traces of life, while the biological results of the in-situ field measurements conducted by NASA's Viking Project landers have remained tantalizingly unclear. However, because their results now strongly suggest that ancient Mars was once a much wetter world, the more recent robot spacecraft missions to the surface of the Red Planet have again raised the question of life there.

A second major approach in exobiology involves conducting experiments in terrestrial laboratories or in laboratories in space that attempt either to simulate the primeval conditions that led to the formation of life on Earth and extrapolate these results to other planetary environments or to study the response of terrestrial organisms under environmental conditions that were found on alien worlds.

In 1924, Aleksandr Ivanovich Oparin (1884–1980), a Russian biochemist, published the book *The Origins of Life on Earth* in which he proposed a theory of chemical evolution that suggested that organic compounds could be produced from simple inorganic molecules and that life on Earth probably originated by this process. (The significant hypothesis of his book

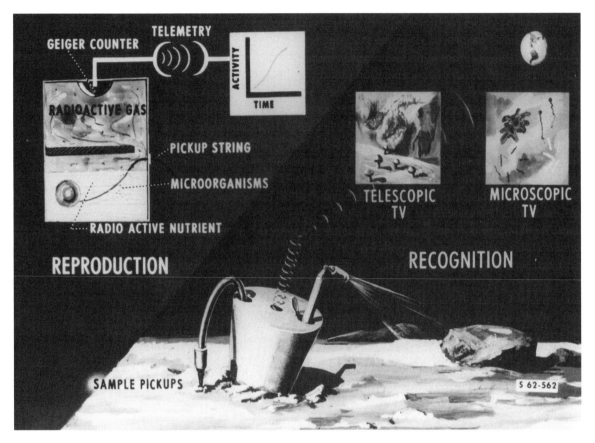

This early space age diagram (ca. 1962) illustrates the basic principles of spacecraft-conducted exobiology. A robot spacecraft might search for life on Mars (or another alien world) by injecting a radioactive nutrient (one containing carbon 14, for example) into one or more soil samples. If the soil sample contains any microscopic life-forms, then the nutrient solution should stimulate metabolic activity and/or growth (reproduction)—resulting in the release of a tiny quantity of radioactive gas, which is easily detected and reported back to Earth. The spacecraft's telescopic television systems could survey the local region for signs of plant and or small animal life. Finally, a high-resolution microscope would allow scientists on Earth to examine scooped up soil samples remotely for signs of tiny living things, such as worms, insects, or possibly the fossilized remains of small, ancient creatures. The Martian equivalent of a clamshell would be an exciting example. *(NASA)*

remained isolated within the former Soviet Union and was not even translated into English until 1938.) A similar theory was also put forward in 1929 by John Burdon Haldane (1892–1964), a British biologist. Unfortunately, the chemical evolution of life hypothesis remained essentially dormant within the scientific community for another two decades. Then, in 1953, at the University of Chicago, American Noble laureate Harold C. Urey (1893–1981) and his former student Stanley L. Miller (1930–   ) performed what many scientists consider as the first modern experiments in exobiology. While investigating the chemical origin of life, Urey and Miller demonstrated that organic molecules could indeed be produced by irradiating a mixture of inorganic molecules. The historic Miller-Urey experiment simulated the Earth's assumed primitive atmosphere by using a gaseous mixture of methane (CH_4), ammonia (NH_3), water vapor (H_2O), and hydrogen (H_2) in a glass flask. A pool of water was kept gently boiling to promote circulation within the mixture, and an electrical discharge (simulating lightning) provided the energy needed to promote chemical reactions. Within days, the mixture changed colors, indicating that more complex, organic molecules had been synthesized out of this primordial "soup" of inorganic materials.

Scientists have used electric discharge equipment (like the laboratory apparatus shown here) to simulate conditions in the primitive atmosphere of Earth that may have produced the basic chemical building blocks of life. Under the influence of primitive planetary environmental conditions, simple organic molecules combined to form larger, more complex organic molecules. Eventually, very large complex molecules assembled into structures that exhibited the key properties of living things: metabolism, respiration, reproduction, and the transfer of genetic information. The most famous version of this type of exobiology experiment is the 1953 Miller–Urey experiment conducted at the University of Chicago by graduate student Stanley L. Miller under the mentorship of his research advisor, Nobel laureate Harold C. Urey. (NASA)

Amino Acid

Amino acid is an acid containing the amino (NH_2) group, a group of molecules considered necessary for life. More than 80 amino acids are presently known, but only some 20 occur naturally in living organisms, where they serve as the building blocks of proteins. On Earth, many microorganisms and plants can synthesize amino acids from simple inorganic compounds. However, terrestrial animals (including human beings) must rely on their diet to provide adequate supplies of amino acids.

Scientists have synthesized amino acids nonbiologically under conditions that simulate those that may have existed on the primitive Earth, followed by the synthesis of most of the biologically important molecules. Amino acids and other biologically significant organic substances have also been found to occur naturally in meteorites and are not considered to have been produced by living organisms.

A third general approach in exobiology involves an attempt to communicate with, or at least to listen for signals from, other intelligent life-forms within the Milky Way galaxy. This effort often is called the search for extraterrestrial intelligence (SETI). Contemporary SETI activities throughout the world have as their principal aim to listen for evidence of extraterrestrial radio signals generated by intelligent alien civilizations. (See chapter 8 for a more detailed discussion of SETI.)

✧ Extraterrestrial Life Chauvinisms

At present, scientists have only one scientific source of information on the emergence of life in a planetary environment—our own planet Earth. They currently believe that all carbon-based terrestrial organisms have descended from a common, single occurrence of the origin of life in the primeval "chemical soup" ocean of an ancient Earth. How can exobiologists project this singular fact to possibly billions of unvisited worlds in the galaxy? As credible scientists, they can only do so with great sense of technical caution. Exobiologists recognize full well that their models of extraterrestrial life-forms and their probabilistic estimates concerning the cosmic prevalence of life on suitable extrasolar planets can easily become prejudiced, or chauvinistic.

Chauvinism is defined as a strongly prejudiced belief in the superiority of one's group. Applied to speculations about extraterrestrial life, this word can take on several distinctive meanings, each heavily influencing any subsequent thought on the subject. Some of the more common forms

SPECTRAL CLASSIFICATION OF STARS

The spectral classification of stars allows astronomers to assign individual stars scientifically useful, quantitative designations. In the 1890s, astronomers at the Harvard College Observatory (HCO) introduced a system of letters that corresponded to different types of stars, based on the stellar spectral lines. The Harvard classification system corresponded roughly to the observable, surface temperatures of these various groups of stars. Astronomers still use this system and classify stars as O (hottest), B, A, F, G, K, or M (coolest). M stars are numerous, long-lived but very dim, while O and B stars are very bright but short-lived and rare.

The Harvard system involves a sequence established in order of decreasing surface temperature, ranging from about 35,000 kelvins for O-type stars to less than 3,500 kelvins for M-type stars. The corresponding colors of the stars in this system are: O (very hot, large blue stars), B (large blue stars), A (blue-white stars), F (white stars), G (yellow stars), K (orange-red stars), and M (red stars).

In 1943, the astronomers William Wilson Morgan (1906–94) and Philip Child Keenan (1908–2000) refined the Harvard classification system by subdividing each lettered spectral classification into 10 subdivisions, denoted by the numbers 0 to 9. By convention within modern astronomy, the hotter the star, the lower this number. Astronomers classify the Sun as a G2 star. This means the Sun is a bit hotter than a G3 star and a bit cooler than a G1 star. For comparison, Betelgeuse is an M2 star and Vega is an A0 star.

Astronomers have also found it helpful to categorize stars by luminosity class. The standard stellar luminosity classes are: Ia for bright supergiants, Ib for supergiants, II for bright giants, III for giants, IV for subgiants, V for main sequence stars (also called dwarfs), VI for subdwarfs, and VII for white dwarfs. This classification scheme is based upon a single spectral property—the width of a star's spectral lines. From astrophysics, scientists know that spectral-line width is sensitive to the density conditions in a star's photosphere. In turn, astronomers can correlate a star's atmospheric density to its luminosity. The use of luminosity classes allows astronomers to distinguish giants from main-sequence stars, supergiants from giants, and so forth. As a main-sequence (dwarf) star, astronomers assign the Sun the stellar luminosity class V.

With a surface temperature of 5,800 kelvins, the full spectral classification of the familiar yellow star that humans call the Sun is G2V. G-star (or solar-system) chauvinism implies, therefore, that life can only originate in a star system like our own—namely, a system containing a single G–spectral-class star.

of extraterrestrial life chauvinisms are: G-star chauvinism, planetary chauvinism, terrestrial chauvinism, chemical chauvinism, oxygen chauvinism, and carbon chauvinism. Although such heavily steeped thinking may not actually be wrong, it is important to realize that it also sets limits, intentionally or unintentionally, on contemporary speculations about life in the universe.

Planetary chauvinism assumes that extraterrestrial life has to develop independently on a particular type of planet, while terrestrial chauvinism stipulates that only "life as scientists know it on Earth" can originate elsewhere in the universe. Chemical chauvinism demands that extraterrestrial life be based on chemical processes, while oxygen chauvinism states that alien worlds must be considered uninhabitable if their atmospheres do not contain oxygen. Finally, carbon chauvinism asserts that extraterrestrial life-forms must be based on carbon chemistry.

These chauvinisms, singularly and collectively, impose tight restrictions on the type of planetary system that scientists suspect might support the rise of living systems elsewhere in the universe—possibly to the level of intelligence. If such restrictions are indeed correct, then the overall search for extraterrestrial intelligence is now being properly focused on Earth-like worlds around Sunlike stars.

If, on the other hand, life is actually quite prevalent and capable of arising in a variety of independent biological scenarios (for example, silicon-based or sulfur-based chemistry), then our contemporary efforts in modeling the cosmic prevalence of life and in trying to describe what "little green men" really look like is somewhat analogous to using the atomic theory of Democritus, the ancient Greek philosopher (ca. 460–ca. 370 B.C.E.), to help describe the inner workings of a modem nuclear-fission reactor.

As scientists continue to explore planetary bodies in this solar system—especially Mars and the interesting larger moons of Jupiter (such as Europa) and Saturn (such as Titan)—they will be able to assess more effectively how valid these extraterrestrial life chauvinisms really are. The discovery of extinct or existing alien life on any (or all) of these neighboring worlds would suggest that the universe is probably teeming with life in a wide variety of previously unimaginable forms.

✧ NASA's Viking Project

The Viking Project was the culmination of an initial series of U.S. missions to explore Mars in the 1960s and 1970s. This series of interplanetary missions began in 1964 with *Mariner 4,* continued with the *Mariner 6* and *7* flyby missions in 1969, and ended to date with the *Mariner 9* orbital mission in 1971 and 1972. Viking robot spacecraft were designed as a complement pair: one to orbit Mars and one to land and operate on the surface of the Red Planet. Reflecting the aerospace engineering philosophy of redundant missions that were characteristic of the 1970s, NASA constructed and launched two identical missions, each consisting of a lander and an orbiter. In many respects, the Viking spacecraft represented the most sophisticated attempt by humans in the 20th century to find life on Mars using robot spacecraft. Chapter 4 discusses contemporary attempts

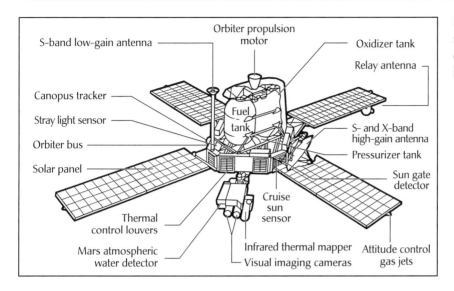

NASA's Viking orbiter spacecraft and its complement of instruments. *(NASA)*

to look for life on Mars with robot spacecraft, many of which build upon the Viking Project's technical legacy.

The Viking Project orbiters carried the following scientific instruments:

1. A pair of cameras with 1,500-millimeter focal length that performed systematic searches for landing sites, then looked at and mapped almost 100 percent of the Martian surface. Cameras onboard the *Viking 1* and *Viking 2* orbiters took more than 51,000 photographs of Mars.
2. A Mars atmospheric water detector that mapped the Martian atmosphere for water vapor and tracked seasonal changes in the amount of water vapor.
3. An infrared thermal mapper that measured the temperatures of the surface, polar caps, and clouds; it also mapped seasonal changes. In addition, although the Viking orbiter radios were not considered scientific instruments, they were used as such. By measuring the distortion of radio signals as these signals traveled from the Viking orbiter spacecraft to Earth, scientists were also able to measure the density of the Martian atmosphere. The ground-based, large radio telescopes of NASA's Deep Space Network (DSN) played a critical role in these efforts.

The Viking Project landers carried the following instruments:

1. The biology instrument, consisting of three separate experiments that were designed to detect evidence of microbial life in the

NASA's Viking lander spacecraft and its complement of instruments. *(NASA)*

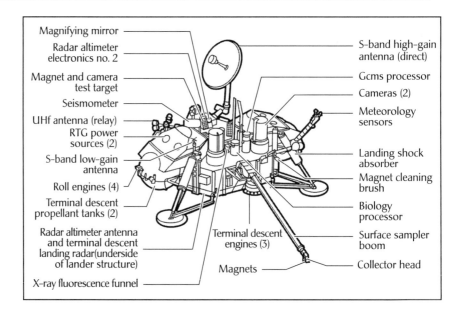

Martian soil. There was always a remote chance that larger life-forms could be present on Mars. But NASA's exobiologists thought then (as they do now) that any native life-forms currently existing on Mars would most likely be microorganisms.

2. A gas chromatograph/mass spectrometer (GCMS) that searched the Martian soil for complex organic molecules. Any such complex organic molecules could be the precursors or the remains of living organisms.

3. An X-ray fluorescence spectrometer that analyzed samples of the Martian soil to determine its elemental composition.

4. A meteorology instrument that measured air temperature and wind speed and direction at the landing sites. These instruments returned the first extraterrestrial weather reports in the history of meteorology.

5. A pair of slow-scan cameras that were mounted about 3-feet (1-m) apart on the top of each lander spacecraft. These cameras provided black-and-white, color, and stereo photographs of the Martian surface.

6. A seismometer that was designed to record any "Marsquakes" that might occur on the Red Planet. Such information would have helped planetary scientists determine the nature of the planet's internal structure. Unfortunately, the seismometer on the *Viking 1* lander did not function after landing, and the instrument on the *Viking 2* lander observed no clear signs of internal (tectonic) activity on the planet.

7. An upper-atmosphere mass spectrometer that conducted its primary measurements as each robot lander plunged through the Martian atmosphere on its way to the landing site. This instrument made the *Viking 1* lander's first important scientific discovery, the presence of nitrogen in the Martian atmosphere.

8. A retarding potential analyzer that measured the Martian ionosphere, again during entry operations.

9. Accelerometers, a stagnation pressure instrument, and a recovery temperature instrument that helped determine the structure of the lower Martian atmosphere as each of the lander spacecraft approached the surface of the planet.

10. A surface sampler boom that employed its collector head to scoop up small quantities of Martian soil to feed the biology, organic-chemistry, and inorganic-chemistry instruments. It also provided clues to the soil's physical properties. Magnets attached to the sampler, for example, provided information on the soil's iron content.

11. In addition to sending scientific data back to scientists on Earth via NASA's Deep Space Network, the lander spacecraft radios also were used to conduct scientific experiments. Physicists were able to refine their estimates of Mars's orbit by measuring the time for radio signals to travel between Mars and Earth. The great accuracy of these radio-wave measurements also allowed scientists to confirm portions of Albert Einstein's general theory of relativity.

Both Viking missions were launched from Cape Canaveral, Florida. *Viking 1* was launched on August 20, 1975, and *Viking 2* on September 9, 1975. The landers were carefully sterilized before launch to prevent contamination of Mars by terrestrial microorganisms. These spacecraft spent nearly a year in transit to the Red Planet. *Viking 1* achieved Mars orbit on June 19, 1976, and *Viking 2* began to orbit Mars on August 7, 1976. The *Viking 1* lander spacecraft accomplished the first soft landing on Mars on July 20, 1976, on the western slope of Chryse Planitia (the Plains of Gold) at 22.46° north latitude, 48.01° west longitude. The *Viking 2* lander touched down successfully on September 3, 1976, at Utopia Planitia (the Plains of Utopia) at 47.96° north latitude, 225.77° west longitude.

The Mars surface science portion of the Viking mission was originally planned to be conducted for approximately 90 days after landing. However, each orbiter and lander pair successfully operated far beyond their design lifetimes. For example, the *Viking 1* orbiter exceeded four years of active flight operations in orbit around Mars. The Viking Project's primary mission ended on November 15, 1976, just 11 days

NASA's Approach to Reducing the Bioload of the Viking Landers

NASA engineers and scientists used a two-fold approach to control the population of "hitchhiking" terrestrial microorganisms that could find their way to the surface of Mars—thereby causing forward contamination of the Red Planet. The first step involved a very careful presterilization cleaning of the lander spacecraft during assembly; the second step involved postassembly heat sterilization.

Engineers and technicians carefully assembled the *Viking 1* and *2* lander spacecraft in Class 100,000 clean rooms. During assembly operations, the technical team conducted thousands of microbial assays. These assays established that the average spore burden was less than 28 per square foot (300 per square meter) and that the total burden of

spores on the lander spacecraft's surface was less than 300,000. In performing the microbiological assays, NASA personnel used the spore-forming microbe *Bacillus subtilis* as the indicator organism because of this microbe's enhanced resistance to heat, radiation, and desiccation.

After the *Viking 1* and *2* landers had been assembled and sealed inside their respective bioshields, the bioload of each lander spacecraft was further reduced by dry heating. Aerospace workers heated the landers to a minimum temperature of 233°F (111.7°C) for about 30 hours. NASA exobiologists estimated that the bioburden of each lander spacecraft was reduced by a factor of 10^4 as a result of this thermal sterilization procedure.

before Mars passed behind the Sun (an astronomical event called a superior conjunction). After conjunction, in mid-December 1976, telemetry and command operations were reestablished, and extended mission operations began.

The *Viking 2* orbiter mission ended on July 25, 1978, due to exhaustion of attitude-control system gas. The *Viking 1* orbiter spacecraft also began to run low on attitude-control system gas, but through careful planning, it was possible to continue collecting scientific data (at a reduced level) for another two years. Finally, with its control gas supply exhausted, the *Viking 1* orbiter's electrical power was commanded off on August 7, 1980.

The last data from the *Viking 2* lander were received on April 11, 1980. The *Viking 1* lander made its final transmission to Earth on November 11, 1982. After more than six months of futile efforts to regain contact with the *Viking 1* lander, NASA mission controllers ended Viking mission operations on May 23, 1983.

With the single exception of the seismic instruments, the entire complement of scientific instruments carried off the Viking Project spacecraft acquired far more data about Mars than ever anticipated. The seismometer on the *Viking 1* lander did not function after touchdown, while the seismometer on the *Viking 2* lander detected only one event that might have been of seismic origin. Nevertheless, the instrument still provided data on surface wind velocity at the Utopia Planitia site (supplementing the meteorology experiment) and also indicated that the Red Planet currently has a very low level of seismicity.

The primary objective of the robot landers was to determine whether life currently exists on Mars. Three of the lander's instruments were capable of detecting microbial life on Mars. In addition, the lander cameras could have photographed any living creatures large enough to be seen with the human eye. These cameras would also have observed growth in organisms such as plants and lichens. Unfortunately, the cameras at both landing sites observed nothing that could be interpreted as living. Although the evidence provided by the *Viking 1* and *2* landers concerning the presence of microbial life is still subject to some debate, today most scientists regard the null results as strongly indicative of the fact that life does not currently exist on Mars—at least at either landing site.

The gas chromatograph/mass spectrometer (GCMS) could have found organic molecules in the soil. (Organic compounds combine carbon, nitrogen, hydrogen, and oxygen.) These compounds are present in all living matter on Earth. The GCMS was programmed to search for heavy organic molecules, those large molecules that contain complex combinations of carbon and hydrogen and are either life precursors or the remains of living systems. To the surprise of exobiologists, the GCMS (which easily detected organic matter in the most barren soils found on Earth) found no trace of any organic molecules in the Martian soil samples scooped up and tested at each landing site.

The biology instrument on each lander spacecraft was the primary device used to search for extraterrestrial life. It was a 1-cubic-foot (0.0286-cubic-meter) box, loaded with the most sophisticated scientific instrumentation yet built and flown in space. The biology instrument actually contained three smaller instruments that examined the Martian soil for evidence of metabolic processes like those used by bacteria, green plants, and animals here on Earth.

The three biology experiments worked flawlessly on each Viking lander. All showed unusual activity in the Martian soil—activity that mimicked life—but exobiologists here on Earth needed time to understand the strange behavior of the Red Planet's soil. Today, according to most scientists who helped analyze these data, it appears that the chemical reactions were not caused by living things.

Furthermore, the immediate release of oxygen, when the Martian soil contacted water vapor in the biology instrument, and the lack of organic compounds in the soil indicated that oxidants were present in both the

Martian soil and the atmosphere. Oxidants, such as peroxides and super-oxides, are oxygen-bearing compounds that break down organic matter and living tissue. Consequently, even if organic compounds evolved on Mars, they would have been quickly destroyed.

Evaluation of the Martian atmosphere and soil has revealed that all the elements essential for life (as known on Earth)—carbon, hydrogen, nitrogen, oxygen, and phosphorus—are also present on the Red Planet. However, exobiologists currently consider the presence of liquid water on a planet's surface as an absolute requirement for the evolution and continued existence of life. The Viking Project discovered ample evidence of Martian water in two of its three phases, namely vapor and solid (ice), and even evidence of large quantities of permafrost. But under current environmental conditions on Mars, it is impossible for water to exist as a liquid on the planet's surface.

Viking spacecraft data indicated that the conditions now occurring on and just below the surface of the Red Planet do not appear adequate for the existence of living (carbon-based) organisms. However, exobiologists, though disappointed in their first serious search for extraterrestrial life, recognize that the case for life sometime in the past history of Mars is still open. Some scientists even cautiously speculate that viable microbial life-forms might still be found in selective, subsurface enclaves where small quantities of liquid water may possibly occur. The search for such ecological niches is one of the major objectives of the numerous NASA robotic missions to Mars in the first decade of this century. (Chapter 4 discusses some of these efforts.)

While the gas chromatograph/mass spectrometer found no sign of organic chemistry at either landing site, the instrument did provide a precise and definitive analysis of the composition of the Martian atmosphere. For example, the GCMS found previously undetected trace elements. The lander spacecraft's X-ray fluorescence spectrometer measured the elemental composition of the Martian soil.

In addition to conducting an automated search for life, the two robot landers continuously monitored weather at the landing sites. The midsummer Martian weather proved repetitious, but in other seasons, the weather varied and became more interesting. Cyclic variations in Martian weather patterns were observed. Atmospheric temperatures at the southern (*Viking 1*) landing site were as high as 6.8°F (-14°C) at midday, while the predawn summer temperature was typically -107°F (-77°C). In contrast, the diurnal temperatures at the northern (*Viking 2*) landing site during the midwinter dust storm varied as little as 39°F (4°C) on some days. The lowest observed predawn temperature was -184°F (-120°C), which is about the frost point of carbon dioxide. A thin layer of water frost covered the ground near the *Viking 2* lander each Martian winter.

The barometric pressure was observed to vary at each landing site on a semiannual basis. This occurred because carbon dioxide (the major constituent of the Martian atmosphere) freezes to form an immense polar cap, alternately at each pole. The carbon dioxide forms a great cover of "snow" and then evaporates (or sublimes) again with the advent of Martian spring in each hemisphere. When the southern cap was largest, the mean daily pressure observed by *Viking 1* lander was as low as 6.8 millibars (680 Pa), while at other times during the Martian year, it was as high as 9.0 millibars (900 Pa). Similarly, the pressures at the *Viking 2* lander site were 7.3 millibars (730 Pa) (full northern cap) and 10.8 millibars (1,080 Pa). For comparison, the sea-level atmospheric pressure on Earth is about 1,000 millibars (101,300 Pa).

Martian surface winds were also generally slower than anticipated. Scientists had expected these winds to reach speeds of a few hundred miles (km) per hour. But neither Viking lander recorded a wind gust in excess of 75 miles per hour (120 km/h), and average speeds were considerably lower.

Photographs of Mars from the Viking landers and orbiters surpassed all expectations in both quantity and quality. Together, the *Viking 1* and *2* landers provided more than 4,500 images, and the *Viking 1* and *2* orbiters more than 51,000. The landers provided the first close-up view of the surface of the Red Planet, while the orbiters mapped almost 100 percent of the Martian surface, including detailed images of many intriguing surface features.

The infrared thermal mapper and the atmospheric water detector onboard the *Viking 1* and *2* orbiters provided essentially daily data. Through these data, it was determined that the residual northern polar ice cap that survives the northern summer is composed of water ice, rather than frozen carbon dioxide (dry ice), as scientists once believed.

Today, after all the Viking Project robot explorers have fallen silent, their data represents a valuable technical heritage that supports the current wave of investigation of Mars with even more sophisticated robot spacecraft. Following in the footsteps of the highly successful Viking spacecraft, new generations of robot explorers now scan and scamper across the surface of the Red Planet hoping to answer the intriguing questions about Mars that still remain, especially in the fields of exobiology and comparative planetology. Is there a remote possibility that life exists in some crevice or biological niche on this mysterious world? Did life once evolve there, only to have vanished millions of years ago? And how did climatic conditions change so radically that great floods of water, which apparently raged over the Martian plains, have now vanished, leaving behind the dry, sterile world found by the Viking Project explorers? Only further exploration during this century, including possibly human expeditions, can solve these intriguing mysteries.

✧ Deep Space Network (DSN)

The majority of NASA's scientific investigations of the solar system are accomplished through the use of robot spacecraft. The Deep Space Network (DSN) provides the two-way communications link that guides and controls these spacecraft and brings back the spectacular planetary images and other important scientific data they collect.

The DSN consists of telecommunications complexes strategically placed on three continents—providing almost continuous contact with scientific spacecraft traveling in deep space as Earth rotates on its axis. The Deep Space Network is the largest and most sensitive scientific telecommunications system in the world. It also performs radio and radar astronomy observations in support of NASA's mission to explore the solar system and the universe. The Jet Propulsion Laboratory (JPL) in Pasadena, California, manages and operates the Deep Space Network for NASA.

The Jet Propulsion Laboratory established the predecessor to the DSN. Under a contract with the U.S. Army in January 1958, the laboratory deployed portable radio tracking stations in Nigeria (Africa), Singapore (Southeast Asia), and California to receive signals from and plot the orbit of *Explorer 1*—the first U.S. satellite to orbit Earth successfully. Later that year (on December 3, 1958), as part of the emergence of the new federal civilian space agency, JPL was transferred from U.S. Army jurisdiction to that of NASA. At the very onset of the U.S. civilian space program, NASA assigned JPL responsibility for the design and execution of lunar and planetary exploration programs by robot spacecraft. Shortly afterward, NASA embraced the concept of the DSN as a separately managed and operated telecommunications facility that would accommodate all deep space missions. This management decision avoided the need for each space flight project to acquire and operate its own specialized telecommunications network.

Today, the DSN features three deep space communications complexes placed approximately 120 degrees apart around the world: at Goldstone in California's Mojave Desert; near Madrid, Spain; and near Canberra, Australia. This global configuration ensures that, as Earth rotates, an antenna is always within sight of a given spacecraft, day and night. Each complex contains up to 10 deep space communication stations equipped with large parabolic reflector antennas.

Every deep space communications complex within the DSN has a 230-foot- (70-m-) diameter antenna. These antennas, the largest and most sensitive in the DSN network, are capable of tracking spacecraft that are more than 10 billion miles (16 billion km) away from Earth. The 41,450-square-feet (3,850-square-meter) surface of the 230-foot- (70-m-) diameter reflector must remain accurate within a fraction of the signal wavelength, meaning that the dimensional precision across the surface is

A view of the 230-foot- (70-m-) diameter antenna of the Canberra Deep Space Communications Complex located outside Canberra, Australia. This facility is one of the three complexes that comprise NASA's deep space network (DSN). The other complexes are located in Goldstone, California, and Madrid, Spain. The national flags representing the three DSN sites appear in the foreground of this image. *(NASA)*

maintained to within 0.4 inch (1 cm). The dish and its mount have a mass of nearly 15.8 million pounds (7.2×10^6 kg).

There is also a 112-foot- (34-m-) diameter high-efficiency antenna at each complex, which incorporates advances in radio frequency antenna design and mechanics. The reflector surface of the 112-foot- (34-m-) diameter antenna is precision-shaped for maximum signal-gathering capability.

The most recent additions to the DSN are several 112-foot (34-m) beam waveguide antennas. On earlier DSN antennas, sensitive electronics were centrally mounted on the hard-to-reach reflector structure, making upgrades and repairs difficult. On beam waveguide antennas, the sensitive electronics are now located in a below-ground pedestal room. Telecommunications engineers bring an incident radio signal from the reflector to this room through a series of precision-machined radio frequency reflective mirrors. Not only does this architecture provide the advantage of easier access for maintenance and electronic equipment enhancements, but the new configuration also accommodates better thermal control of critical electronic components. Furthermore, engineers can place more electronics in the antenna to support operation at multiple frequencies. Three of these new 112-foot (34-m) beam waveguide antennas have been constructed at the Goldstone, California, complex, along with one each at the Canberra and Madrid complexes.

There is also one 85-foot- (26-m-) diameter antenna at each complex for tracking Earth-orbiting satellites, which travel primarily in orbits of 100 miles (160 km) to 620 miles (1,000 km) above Earth. The two-axis astronomical mount allows these antennas to point low on the horizon to acquire (pick up) fast-moving satellites as soon as they come into view. The agile 85-foot- (26-m-) diameter antennas can track (slew) at up to three degrees per second. Finally, each complex also has one 36-foot- (11-m-) diameter antenna to support a series of international Earth-orbiting missions involving very long baseline interferometry.

All of the antennas in the DSN network communicate directly with the Deep Space Operations Center (DSOC) at JPL in Pasadena, California. The DSOC staff directs and monitors operations, transmits commands, and oversees the quality of spacecraft telemetry and navigation data delivered to network users. In addition to the DSN complexes and the operations center, a ground communications facility provides communications that link the three complexes to the operations center at JPL, to spaceflight control centers in the United States and overseas, and to scientists around the world. Voice and data communications traffic between various locations is sent via landlines, submarine cable, microwave links, and communications satellites.

The Deep Space Network's radio link to scientific robot spacecraft is basically the same as other point-to-point microwave communications systems, except for the very long distances involved and the very low radio frequency signal strength received from the robot spacecraft. The total signal power arriving at a network antenna from a typical robot spacecraft encounter among the outer planets can be 20 billion times weaker than the power level in a modern digital wristwatch battery.

The extreme weakness of this radio frequency signals results from restrictions placed on the size, mass, and power supply of a particular spacecraft by the payload volume and mass-lifting limitations of its launch vehicle. Consequently, the design of the radio link is the result of engineering trade-offs between spacecraft transmitter power and antenna diameter and the signal sensitivity that engineers can build into the ground receiving system.

Typically, a spacecraft signal is limited to 20 watts, or about the same amount of power required to light the bulb in a refrigerator. When the spacecraft's transmitted radio signal arrives at Earth—from, for example, the neighborhood of Saturn—it has spread over an area with a diameter equal to about 1,000 Earth diameters. (Earth has an equatorial diameter of 7,928 miles [12,756 km]). As a result, the ground antenna is able to receive only a very small part of the signal power, which is also degraded by background radio noise, or static.

Radio noise is radiated naturally from nearly all objects in the universe, including Earth and the Sun. Noise is also inherently generated in all electronic systems including the DSN's own detectors. Since noise will always be amplified along with the signal, the ability of the ground receiving system to separate noise from the signal is critical. The DSN uses state-of-the-art, low-noise receivers and telemetry coding techniques to create unequaled sensitivity and efficiency.

Telemetry is basically the process of making measurements at one point and transmitting the data to a distant location for evaluation and use. A robot spacecraft sends telemetry to Earth by modulating data onto its communications downlink. Telemetry includes state-of-health data about the spacecraft's subsystems and science data from its instruments. A typical scientific spacecraft transmits its data in binary code, using only the symbols 1 and 0. The spacecraft's data-handling subsystem (telemetry system) organizes and encodes these data for efficient transmission to ground stations back on Earth. The ground stations have radio antennas and specialized electronic equipment to detect the individual bits, decode the data stream, and format the information for subsequent transmission to the data user (usually a team of scientists).

Data transmission from a robot spacecraft can be disturbed by noise from various sources that interferes with the decoding process. If there is a high signal-to-noise ratio, the number of decoding errors will be low. But if the signal-to-noise ratio is low, then an excessive number of bit errors can occur. When a particular transmission encounters a large number of bit errors, mission controllers will often command the spacecraft's telemetry system to reduce the data transmission rate (measured in bits per second) to give the decoder (at the ground station) more time to determine the value of each bit.

To help solve the noise problem, a spacecraft's telemetry system might feed additional or redundant data into the data stream, which additional data are then used to detect and correct bit errors after transmission. The information theory equations used by telemetry analysts in data evaluation are sufficiently detailed to allow the detection and correction of individual and multiple bit errors. After correction, the redundant digits are eliminated from the data, leaving a valuable sequence of information for delivery to the data user.

Error detecting and encoding techniques can increase the data rate many times over transmissions that are not coded for error detection. DSN coding techniques have the capability of reducing transmission errors in spacecraft science information to less than one in a million.

Telemetry is a two-way process, having a downlink as well as an uplink. Robot spacecraft use the downlink to send scientific data back to Earth, while mission controllers on Earth use the uplink to send commands, computer software, and other crucial data to the spacecraft. The uplink portion of the telecommunications process allows human beings to guide spacecraft on their planned missions, as well as to enhance mission objectives through such important activities as upgrading a spacecraft's onboard software while the robot explorer is traveling through interplanetary space. When large distances are involved, human supervision and guidance is limited to non-real time interactions with the robot spacecraft. That is why deep space robots must possess high levels of machine intelligence and autonomy.

Data collected by the DSN is also very important in precisely determining a spacecraft's location and trajectory. Teams of human beings (called the mission navigators) use these tracking data to plan all the maneuvers necessary to ensure that a particular scientific spacecraft is properly configured and is at the right place (in space) to collect its important scientific data. Tracking data produced by the DSN let mission controllers know the location of a spacecraft that is billions of miles (kilometers) away from Earth to an accuracy of just a few feet (meters).

NASA's Deep Space Network is also a multifaceted science instrument that scientists can use to improve their knowledge of the solar system and the universe. For example, scientists use the large antennas and sensitive electronic instruments of the DSN to perform experiments in radio astronomy, radar astronomy, and radio science. The DSN antennas collect information from radio signals emitted or reflected by natural celestial sources. Such DSN-acquired radio frequency data are compiled and analyzed by scientists in a variety of disciplines, including astrophysics, radio astronomy, planetary astronomy, radar astronomy, Earth science, gravitational physics, and relativity physics.

In its role as a science instrument, the DSN provides the information that is needed to: select landing sites for space missions; determine the composition of the atmospheres and/or the surfaces of the planets and their moons; search for biogenic elements in the interstellar space; study the process of star formation; image asteroids; investigate comets, especially their nuclei and comas; search the permanently shadowed regions of the Moon and Mercury for the presence of water ice; and confirm Albert Einstein's theory of general relativity.

The DSN radio science system performs experiments that allow scientists to characterize the atmospheres and ionospheres of planets, determine the compositions of planetary surfaces and rings, look through the solar corona, and determine the mass of planets, moons, and asteroids. It accomplishes this by precisely measuring the small changes that take place in a spacecraft's telemetry signal as the radio waves are scattered, refracted, or absorbed by particles and gases near celestial objects within the solar system. The DSN makes its facilities available to qualified scientists as long as the research activities do not interfere with spacecraft mission support.

✧ NASA's Origins Program

Thousands of years ago, on a small rocky planet orbiting a modest star in an ordinary spiral galaxy, our prehistoric ancestors looked up and wondered about their place between Earth and sky. In the 21st century, human beings ask the same profound questions: How did the universe begin and evolve? How did we get here? Where are we going? Are we alone?

After only the blink of an eye in cosmic time, humans are beginning to answer some of these questions within the framework of science. Space probes and space-based and ground-based astronomical observatories have played a central role in this process of discovery. NASA's Origins Program involves a series of space missions (present and future) that are intended to help scientists answer these age-old astronomical questions. NASA's Origins-themed missions include the *Spitzer Space Telescope (SST)*, the *James Webb Space Telescope (JWST)* (previously called the *Next Generation Space Telescope*), the *Kepler* spacecraft, the *Space Interferometry Mission (SIM)*, *Terrestrial Planet Finder (TPF)*, the *Single Aperture Far-Infrared Observatory (SAFIR)*, the *Life Finder* and the *Planet Imager*. Earth-based observations, using the Keck Telescopes on Mauna Kea, Hawaii, and the Large Binocular Telescope Interferometer (LBTI) on Mount Graham, Arizona, are also part of the current program. Taken collectively, these powerful astronomical tools represent the sophisticated remote sensing capabilities that will allow astronomers to detect Earth-sized planets around distant stars and determine which are habitable or even inhabited. As incredible as that statement may first appear, the present generation of scientists includes the individuals who, in all likelihood, will discover the first Earth-like planet and determine whether or not there is life on it. This anticipated event would be an amazing moment in exobiology and astronomy.

The central principle of NASA's Origins Program mission architecture has been that each major "planet quest" mission builds on the scientific and technological legacy of previous missions while providing new capabilities for the future. In this way, the complex challenges of the extrasolar planet-hunting theme can be achieved with reasonable cost and acceptable risk. For example, the techniques of interferometry developed for the Keck Interferometer and the *Space Interferometry Mission (SIM)* along with the infrared detector technology from the *Spitzer Space Telescope (SST)* and the large optics technology needed for the *James Webb Space Telescope (JWST)* will enable the Terrestrial Planet Finder mission to search out and characterize habitable planets. This chapter provides currently capabilities in the hunt for extrasolar planets. Chapter 7 discusses how the search for extrasolar planets will be enhanced greatly by missions planned for operation in the next decade or so.

For example, will a launch now scheduled for November 2008, NASA's *Kepler* spacecraft, provide valuable planetary system statistics? This promising mission exemplifies the varied observational approaches the overall Origins Program is attempting to embrace. The dynamic state of this emergent scientific field (that is, extrasolar planet hunting) suggests strongly that the program undertaken to achieve the Origins Program goals must remain flexible and must adapt to and make use of evolving technical and scientific knowledge and capability.

Support for Origins Program missions is provided by two key science centers: the Michelson Science Center (MSC) and the Space Telescope Science Institute (STScI). MSC is a science operations and analysis service that is sponsored by the Origins Program theme and operated by the California Institute of Technology. The MSC facilitates timely and successful execution of projects pursuing the discovery and characterization of extrasolar planetary systems and terrestrial planets. STScI is operated by the Association of Universities for Research in Astronomy Inc. for NASA, under contract with NASA's Goddard Space Flight Center in Greenbelt, Maryland.

Even as scientists and engineers work to develop the missions for this decade and the next decade, they must also begin to envision where their explorations will lead them afterward—since developing the needed advanced space technologies can easily take a decade or more before they are ready to be successfully applied. For example, beyond the *Terrestrial Planet Finder* (TPF), scientific attention should turn to detailed studies of any indications of life found on the planets that TPF discovers. This will require a still-more-capable spectroscopic mission, called Life Finder (LF) mission, which will probe the infrared spectrum with great sensitivity and resolution. Anticipated follow-on space exploration missions, therefore, call for advanced investigations in galaxy and planetary system formation and cosmology that require a high-resolution IR telescope such as a proposed 26-foot- (8-m-) diameter space-based telescope, called the *Single Aperture Far-Infrared Observatory (SAFIR)*.

In a decades-long program of organized investigation, the suggested *Single Aperture Far-Infrared Observatory (SAFIR)* might launch and operate between *TPF* and *LF*—carrying out its own science program while leading to the 82-foot- (25-m-) diameter class spaced-based telescopes needed for *Life Finder (LF)*. The technology developed for such a mission might also be used as a building block for a 1-mile- (1.6-km-) baseline interferometer used at far-infrared wavelengths for cosmological studies. Investigations in distribution of matter in the universe (including dark matter) will require a large-scale UV/optical observatory that will build on the technology developments of the *James Webb Space Telescope (JWST)* and of the *Space Interferometry Mission (SIM)*, and pave the way for more challenging UV/optical telescopes of the future. The astronomical search for scientific answers to the key questions, which form the theme of the Origins Program, provides an inspiring and far-reaching vision that challenges both scientists and engineers.

One of the most exciting parts of NASA's Origins Program is exploring the diversity of other worlds and searching for those that might harbor life. During the past three decades, scientists have used both ground- and space-based facilities to look inside the cosmic nurseries where stars and planets are born. Parallel studies conducted in the solar system with planetary probes and of meteorites have revealed clues to the processes that shaped the early evolution of our own planetary system. An overarching goal of space science in this century is to connect what scientists observe elsewhere in the universe with objects and phenomena in this solar system. Scientists now have strong evidence, based on the telltale wobbles measured for more than 100 nearby stars, that they are orbited by otherwise unseen planets. One remarkable star, Upsilon Andromedae, shows evidence for three giant-planet companions.

Thus far, however, most newly discovered planetary systems around alien stars are quite unlike our own solar system. The masses of the extrasolar planets span a broad range from one-eighth to more than 10 Jupiter masses. Many of these planets (often called *hot Jupiters*) are surprisingly close to their parent stars, and a great many of these large extrasolar planets have eccentric orbits. Extreme proximity and eccentricity are two characteristics not seen in the giant planets of this solar system. Although planets in solar systems like our own are only now becoming detectable with the techniques used to discover the new planets, the lack of a close analog to this solar system and the striking variety of the detected systems raises a fascinating question: Is our solar system of a rare (or even unique) type?

In tandem, astronomers have now identified the basic stages of star formation. The process begins in the dense cores of cold gas clouds (socalled molecular clouds) that are on the verge of gravitational collapse. It continues with the formation of protostars, infant stellar objects with gasrich, dusty circumstellar disks that evolve into adolescent "main-sequence" stars. Tenuous disks of ice and dust that remain after most of the disk gas has dispersed often surround these more mature stars. It is in the context of these last stages of star formation that planets are born. Scientists have now found many extrasolar planets. Most are unlike those in our own solar system. But might there be near-twins of our solar system as well? Are there Earth-like planets? What are their characteristics? Could they support life? Do some actually show signs of past or present life?

✧ Spitzer Space Telescope (SST)

The *Spitzer Space Telescope (SST)* is the final spacecraft in NASA's Great Observatories program—a family of four orbiting observatories each studying the universe in a different portion of the electromagnetic spectrum. This orbiting observatory—previously called the *Space Infrared Telescope Facility (SIRTF)*—consists of a 2.8-foot- (0.85-m-) diameter telescope and three cryogenically cooled science instruments. NASA renamed this space-based infrared telescope in honor of the U.S. astronomer Lyman Spitzer, Jr. (1914–97).

The SST represents the most powerful and sensitive infrared telescope ever launched. The orbiting facility obtains images and spectra of celestial objects at infrared radiation wavelengths between 3 and 180 micrometers (μm)—an important spectral region of observation that is mostly unavailable to ground-based telescopes because of the blocking influence of Earth's atmosphere. Following a successful launch on August 25, 2003, from Cape Canaveral Air Force Station by an expendable Delta rocket, the

This is an artist's rendering of NASA's *Spitzer Space Telescope* in orbit against an infrared (100-micrometer-wavelength) sky. *(NASA/JPL–Caltech)*

2,094-pound- (950-kg-) mass observatory traveled to an Earth-trailing heliocentric orbit. Engineers and mission planners selected this operating orbit to allow the telescope instruments to cool rapidly with a minimum expenditure of onboard cryogen. With a planned mission lifetime in excess of five years, *SST* has taken its place alongside NASA's other great orbiting astronomical observatories and is now collecting high-resolution infrared data that help scientists better understand how galaxies, stars, and planets form and develop.

One major engineering breakthrough with the *Spitzer Space Telescope* was the clever choice of orbit. Instead of orbiting Earth itself, the observatory now trails behind Earth as the planet orbits the Sun. The spacecraft drifts slowly away from Earth into deep space, circling the Sun at a distance of one astronomical unit, which is the mean Earth-Sun distance of approximately 93 million miles (or 150 million km). The telescope is drifting away from Earth at about 1/10th of one astronomical unit per year. This unique orbital trajectory keeps the observatory away from much of Earth's heat, which can reach -10°F (-23°C), or 250 K, for satellites and spacecraft in more conventional near-Earth orbits. The *Spitzer Space Telescope* operates in a more benign thermal environment for infrared telescopes—about -397°F (-238°C), or 35 K. With this innovative design approach, engineers have allowed natural radiation heat-transfer processes to the frigid deep space environment to assist in keeping the observatory properly chilled. Furthermore, the *SST*'s Earth-trailing orbit protects the observatory from Earth's radiation belts. This significantly reduces the harmful effects of ionizing radiation on the observatory's extremely sensitive infrared radiation detectors.

The infrared energy collected by the observatory is being examined and recorded by three main science instruments: an infrared array camera, an infrared spectrograph, and a multiband imaging photometer. The infrared array camera supports imaging at near- and mid-infrared wavelengths. Astronomers use this general-purpose camera for a wide variety of science research programs. The infrared spectrograph allows for both high- and low-resolution spectroscopy at mid-infrared wavelengths. Similar to an optical spectrometer, the infrared spectrograph spreads incoming infrared radiation into its constituent wavelengths. Scientists then scrutinize these infrared spectra for emission and absorption lines—the telltale fingerprints of atoms and molecules. The *Spitzer Space Telescope*'s spectrometer has no moving parts. Finally, the multiband imaging photometer provides imaging and limited spectroscopic data at far-infrared wavelengths. The only moving part in the imaging photometer is a scan mirror for mapping large areas of the sky efficiently.

The *Spitzer Space Telescope*'s powerful combination of highly sensitive detectors and long lifetime allows astronomers to view objects and phenomena that have managed to elude them when they used other observing instruments and astronomical methods. Because of its unique and efficient thermal design, the spacecraft carries only 95 gallons (360 liters) of expendable liquid helium cryogen to cool its sensitive infrared instruments. Cryogen depletion has severely limited the useful lifetime of previous infrared telescopes deployed in space. NASA mission planners estimate that *Spitzer's* cryogen supply is sufficient to provide cooling for the infrared observatory's instrument for about five years of operation. The observatory uses the vapor from the boil-off of its cryogen to cool the infrared telescope assembly down to its optimal operating temperature of -450°F (-268°C), or 5.5 K.

The vast majority of the telescope's observing time is available to the general scientific community through peer-reviewed proposals. The observatory's final design was driven by the goal of making major scientific contributions in the following major research areas: formation of planets and stars (including planetary debris disks and brown dwarf surveys); origin of energetic galaxies and quasars; distribution of matter and galaxies (including galactic halos and missing mass issue); and the formation and evolution of galaxies (including protogalaxies).

Protoplanetary and planetary debris discs are flattened discs of dust that surround many stars. Protoplanetary discs include large amounts of gas and are presumed to be planetary systems in the making. Planetary debris discs have most of their gas depleted and represent a more mature planetary system. The remaining dust disc may include gaps indicative of fledgling planetary bodies. By observing dust discs around stars at various ages, the *Spitzer Space Telescope* can trace the dynamics and chemical history of evolving planetary systems and provide statistical evidence of planetary system formation.

Brown dwarfs are curious infrared objects that do not possess enough mass to gravitationally contract to the point of igniting nuclear fusion reactions in their cores—in the same way that powers true stars. Astronomers consequently call brown dwarfs failed stars. Brown dwarfs are larger and warmer than the planets found in the solar system. At one point, brown dwarfs were considered just a theory, but astronomers have now begun to detect these long-sought objects. High-resolution infrared telescopes, like the *Spitzer Space Telescope,* play a major role in the contemporary search. If brown dwarfs prove to be numerous enough, then they may represent an appreciable fraction of the elusive dark matter or missing mass issue that is now challenging scientists.

Many galaxies emit more radiation in the infrared portion of the spectrum than in all the other wavelength regions combined. These ultraluminous infrared galaxies could be powered by intense bursts of star formation, stimulated by colliding galaxies or by central black holes. The

Spitzer Space Telescope can trace the origins and evolution of ultraluminous infrared galaxies out to cosmological distances (that is, billions of light-years away).

The *Spitzer Space Telescope* is examining galaxies at the cosmic fringe. These objects are so remote that the radiation they once emitted has taken billions of years to reach Earth. A consequence of an expanding universe, these faraway galaxies are receding from Earth so rapidly that most of their optical and ultraviolet light has red-shifted (Doppler effect) into the infrared portion of the spectrum. The *Spitzer Space Telescope* is examining some of these first stars and galaxies to provide scientists with new clues into the character of the infant universe.

Apart from these important research areas in astronomy, the *Spitzer Space Telescope's* near-infrared instrument can peer through obscuring dust, which cocoons newborn stars, both in the nearby universe and also in the center of the Milky Way galaxy. As in the past history of astronomy, whenever there is a giant leap in observational capability, like the *Spitzer Space Telescope* provides in infrared astronomy, there will also be a large number of astronomical surprises and serendipitous discoveries of unanticipated phenomena.

✧ Keck Interferometer

The world's largest ground-based telescopes for optical and infrared astronomy are the twin 32.8-foot- (10-m-) diameter Keck Telescopes at the 13,796-foot (4,206-m) summit of Mauna Kea, a dormant volcano on the island of Hawaii. Each telescope stands eight stories tall and weighs 300 tons, yet operates with nanometer precision. At the heart of each Keck Telescope is a revolutionary primary mirror. Each primary mirror, which is 32.8 feet (10 m) in diameter, is composed of 36 hexagonal segments that work together as a single piece of reflective glass.

Made possible by grants from the W. M. Keck Foundation, the Keck Observatory is operated by the California Association for Research in Astronomy (CARA), whose board of directors includes representatives from the California Institute of Technology (Caltech) and the University of California. In 1996, NASA joined as a partner in the observatory. The Keck I telescope began science observations in May 1993; Keck II began science observations in October 1996.

NASA became involved in the Keck Observatory in the 1990s, primarily to support the agency's interests in detecting and observing planets in other star systems. NASA's astronomers proposed using the two Keck telescopes as an interferometer to conduct these observations. Key science programs for the Keck Interferometer (KI), which are directed at

INTERFEROMETER

The interferometer is an instrument that achieves high angular resolution by combining signals from at least two widely separated telescopes (*optical interferometer*) or a widely separated antenna array (*radio interferometer*). Radio interferometers are one of the basic instruments of radio astronomy. In principle, the interferometer produces and measures interference fringes from two or more coherent wave trains from the same source. These instruments are used to measure wavelengths, to measure the angular width of sources, to determine the angular position of sources (as in satellite tracking), and for many other scientific purposes.

Origins Program objectives, include searching for other planetary systems from their astrometric signature and emitted light, as well as characterizing the environment around nearby stars. The Keck Interferometer has several backend instruments, allowing astronomers to make a variety of observations in the optical and near-infrared portion of the electromagnetic spectrum. On the 279-foot (85-m) Keck-Keck baseline, the interferometer has a spatial resolution of 5 milliarcseconds (mas) when observing at the near infrared wavelength of 2.2 micrometers (μm) and 24 milliarcseconds when observing at the thermal infrared wavelength of 10 micrometers.

The long-range goal of NASA's Origins Program is to discover and characterize Earth-like planets. While this challenging task ultimately requires the use of space-based interferometers—like NASA's planned *Space Interferometry Mission (SIM)* and Terrestrial Planet Finder-Interferometer (TPF-I)—the Keck Interferometer (KI) is now playing an important pathfinder role. Using our home solar system as a model, scientists anticipate that other planetary systems will also have some amount of zodiacal dust (interplanetary dust in the plane of the solar system) around the parent star. Large amounts of this exozodiacal dust can obscure the signature of a planet, making detection far more difficult. So, NASA scientists are using the KI to measure the exozodiacal dust around nearby stars to a level equal to 10 times that of the zodiacal dust emissions in this solar system. The KI uses a nulling technique at 10 micrometers wavelength to null the light from the parent star so that emission from any exozodiacal dust around the target star can be detected. By characterizing the exozodiacal dust around potential targets for the planned TPF mission, NASA scientists are now helping to optimize the design of the TPF mission's instruments, refine the proposed operational data collection scenario, and distill the target list of candidate Earth-like planets.

Astronomers recognize that the direct detection of a "cold Jupiter" can only be accomplished using space-based instruments. A *cold Jupiter* is a massive extrasolar planet (as large or greater than Jupiter in size), that orbits its parent star at some appreciable distance—perhaps five or more astronomical units (AU). In contrast, a *hot Jupiter* is a massive extrasolar planet that has an orbital distance of less than 0.3 AU from its parent star. For comparison, in this solar system, the planet Mercury (which is not classified as a hot Jupiter) orbits the Sun at a distance of about 0.4 AU. The hot Jupiter, extrasolar planets are so close to their parent stars that their surface temperatures are generally quite high, perhaps 1,160°F (627°C) or hotter. Using the sensitive instruments, such as the Keck Interferometer, scientists plan to detect hot Jupiter planets orbiting around nearby stars by virtue of their direct infrared emissions. By applying multicolor phase difference interferometry, the KI provides the capability of directly detecting the radiated (infrared) light from any such massive extrasolar planets that orbit at a separation of 0.15 AU from their parent stars. Astronomers are using the KI in this planet-hunting search to a distance of about 32.6 light-years. The KI's planet detection approach complements the high-precision radial velocity (Doppler) technique that astronomers employed in their initial efforts to detect such massive extrasolar planets. Phase difference interferometry also provides astronomers the opportunity to perform unambiguous mass determinations and validations of atmospheric models.

✧ Large Binocular Telescope Interferometer

The Large Binocular Telescope Interferometer (LBTI) involves a collaborative effort among the Italian astronomical community (represented by the Instituto Nazionale di Astrofisica [INAF]), the University of Arizona, and other academic and research partners—including a consortium of German institutes and observatories called LBT Beteiligungsgesellschaft (LBTB). The LBTI is capable of directly detecting giant planets beyond humans' solar system. The system consists of two 27.6-feet- (8.4-m-) diameter primary mirror telescopes located side by side on Mount Graham, Arizona. This arrangement produces a collecting area that is equivalent to a 39-foot (11.8-m) circular aperture. The two telescopes are connected to form an infrared radiometer with a maximum baseline of 75 feet (22.8 m). The astronomical facility achieved "first light" on October 12, 2005.

Because of its unique geometry and relatively direct optical path, the LBTI provides science capabilities that are different from other interferometers. For example, LBTI provides high-resolution images of many faint objects over a wide field-of-view, including distant galaxies with 10 times the resolution of the *Hubble Space Telescope (HST)*. Of special interest to extrasolar planet hunters is the fact that nulling techniques allow the LBTI to study emissions from faint dust clouds around other stars. Since these dust clouds reflect light and give off heat (thermal radiation), they can interfere with the search for extrasolar planets. So by helping scientists characterize these dust cloud emissions in nearby star systems, the LBTI is supplying very useful data to the NASA scientists and engineers, who are now responsible for the design and development of advanced planet-hunting space missions, such as the Terrestrial Planet Finder (TPF). (Chapter 7 provides additional discussion on this important topic.)

The Hunt for Life on Mars

As a result of NASA's Viking Project in the 1970s, scientists confirmed that the atmosphere of Mars is primarily carbon dioxide (CO_2). Nitrogen, argon, and oxygen are present in small percentages, along with trace amounts of neon, xenon, and krypton. The Martian atmosphere contains only a wisp of water (about 1/1000th as much as found in Earth's atmosphere). But even this tiny amount can condense and form clouds that ride high in the Martian atmosphere or form patches of morning fog in valleys. There is also evidence that Mars had a much denser atmosphere in the past—one that was capable of permitting liquid water to flow on the planet's surface. Physical features resembling riverbeds, canyons and gorges, shorelines, and even islands hint that large rivers and maybe even small seas once existed on the Red Planet.

Among the many scientific discoveries about Mars that were made as a result of early space age exploration, exobiologists regard the possible presence of liquid water on Mars—either in its ancient past or presently preserved somewhere beneath the surface of the planet—as the most important. These scientists consider water as the key in their search for extraterrestrial life because everywhere they find water on Earth, they also find life. So, if Mars once had liquid water or still does today (in some special subsurface niche), there arises a very compelling argument that at least microscopic life could have emerged on the Red Planet.

In the mid-1990s, stimulated by the exciting possibility of life on Mars (extinct or perhaps still existing there today), NASA and other space-exploration organizations launched a variety of robot spacecraft to accomplish more-focused scientific investigations of the Red Planet. Beginning in 1996, some of these missions have proven highly successful, while others have ended as disappointing failures.

In December 2004, NASA planners published a strategic roadmap that defines the civilian space agency's space exploration objectives for the planet Mars, extending to the year 2035 and beyond. The central scientific theme of the focused exploration effort is simply "Follow the water." NASA management views sequenced exploration by a series of sophisticated robot spacecraft as paving the way for larger-scale human exploration missions.

NASA is conducting robotic exploration of Mars to search for evidence of life, to understand the history of the solar system, and to prepare the way for future missions to Mars by human explorers. After acquiring adequate knowledge about the Red Planet through versatile and progressively more-complex robot spacecraft missions, NASA plans to send the first human-crewed mission to Mars sometime around 2035. The discovery of existing microbial life on the Red Planet and concerns about planetary contamination (forward or back) could modify the schedule for human expeditions to the surface of the planet. Robot spacecraft might also discover fossil records on Mars that provide clear evidence of extinct creatures (including microbial life-forms) that once inhabited the Red Planet. This type of momentous discovery would also alter the current, multidecade exploration roadmap.

✧ Mars Pathfinder Mission

NASA launched the *Mars Pathfinder* spacecraft to the Red Planet using a Delta II expendable launch vehicle on December 4, 1996. This mission, previously called the Mars Environmental Survey (or MESUR) Pathfinder, had the primary objective of demonstrating innovative technology for delivering an instrumented lander and free-ranging robotic rover to the Martian surface. The *Mars Pathfinder* not only accomplished this primary mission but also returned an unprecedented amount of data, operating well beyond the anticipated design life.

Mars Pathfinder used an innovative landing method that involved a direct entry into the Martian atmosphere, assisted by a parachute to slow its descent through the planet's atmosphere and then a system of large airbags to cushion the impact of landing. From its airbag-protected bounce and roll landing on July 4, 1997, until the final data transmission on September 27, the robotic lander/rover team returned numerous close-up images of Mars and chemical analyses of various rocks and soil found in the vicinity of the landing site.

The landing site was at 19.33 N, 33.55 W, in the Ares Vallis region of Mars, a large outwash plain near Chryse Planitia (the Plains of Gold), where the *Viking 1* lander had successfully touched down on July 20, 1976. Planetary geologists speculate that this region is one of the largest outflow channels on Mars—the result of a huge ancient flood that occurred during a short period of time and flowed into the Martian northern lowlands.

The lander, renamed by NASA the Carl Sagan Memorial Station, first transmitted engineering and science data collected during atmospheric entry and landing. The U.S. astronomer Carl Edward Sagan (1934–96) popularized astronomy and astrophysics and wrote extensively about the possibility of extraterrestrial life.

Just after arrival on the surface, the lander's imaging system (which was on a pop-up mast) obtained views of the rover and the immediate surroundings. These images were transmitted back to Earth to assist the human flight team in planning the robot rover's operations on the surface of Mars. After some initial maneuvering to clear an airbag out of the way, the lander deployed the ramps for the rover. The 23.3-pound (10.6-kg) minirover had been stowed against one of the lander's petals. Once commanded from Earth, the tiny robot explorer came to life and rolled onto the Martian surface. Following rover deployment, the bulk of the lander's remaining tasks were to support the rover by imaging rover operations and relaying data from the rover back to Earth. Solar cells on the lander's three petals, in combination with rechargeable batteries, powered the lander spacecraft, which also was equipped with a meteorology station.

The rover, renamed *Sojourner* (after the African-American civil-rights crusader Sojourner Truth), was a six-wheeled vehicle that was teleoperated (that is, driven over great distances by remote control) by personnel at the Jet Propulsion Laboratory in Pasadena, California. The rover's human controllers used images obtained by both the rover and the lander systems. Teleoperation at interplanetary distances required that the rover be capable of some semiautonomous operation, since the time delay of the signals averaged between 10 and 15 minutes, depending on the relative positions of Earth and Mars.

For example, the rover had a hazard avoidance system, and surface movement was performed very slowly. The small rover was 11 inches (28 cm) high, 24.8 inches (63 cm) long, and 18.9 inches (48 cm) wide with a ground clearance of 5 inches (13 cm). While stowed in the lander, the rover had a height of just 7.1 inches (18 cm). However, after deployment on the Martian surface, the rover extended to its full height and rolled down a deployment ramp. The relatively far-traveling little rover received its supply of electrical energy from its 2.2-square-foot (0.2-m^2) array of solar cells. Several nonrechargeable batteries provided backup power.

The rover was equipped with a black-and-white imaging system. This system provided views of the lander, the surrounding Martian terrain, and even the rover's own wheel tracks that helped scientists estimate soil properties. An alpha particle X-ray spectrometer (APXS) onboard the rover was used to assess the composition of Martian rocks and soil.

Both the lander and the rover outlived their design lives—the lander by nearly three times and the rover by 12 times. Data from this very successful lander/rover surface mission suggest that ancient Mars was once warm and wet, stimulating further scientific and popular interest in the intriguing question of whether life could have emerged on the planet when it had liquid water on the surface and a thicker atmosphere.

✧ Mars Global Surveyor Mission

NASA launched the *Mars Global Surveyor (MGS)* spacecraft from Cape Canaveral Air Force Station, Florida, on November 7, 1996, using a Delta II expendable launch vehicle. The safe arrival of this robot spacecraft at Mars on September 12, 1997, represented the first successful mission to the Red Planet in two decades. MGS was designed as a rapid, low-cost recovery of the lost *Mars Observer (MO)* spacecraft and its major scientific mission objectives.

After a year and a half of trimming its orbit from a looping ellipse to a circular track around the planet, the spacecraft began its primary mapping mission in March 1999. Using a high-resolution camera, the *MGS* spacecraft observed the planet from a low-altitude, nearly polar orbit over the course of one complete Martian year, the equivalent of nearly two Earth years. Completing its primary mission on January 31, 2001, the spacecraft entered an extended mission phase. NASA lost contact with the *Mars Global Surveyor* in November 2006. This spacecraft had operated longer at Mars than any other spacecraft in history and for more than four times as long as the prime mission originally planned.

The *MGS* science instruments included a high-resolution camera, a thermal emission spectrometer, a laser altimeter, and a magnetometer/electron reflectometer. With these instruments, the spacecraft successfully studied the entire Martian surface, atmosphere, and interior, returning an enormous amount of valuable scientific data in the process. Among the key scientific findings of this mission are high-resolution images of gullies and debris flow features that suggest there may be current sources of liquid water, similar to an aquifer, at or near the surface of the planet.

Magnetometer readings indicate the Martian magnetic field is not globally generated in the planet's core but appears to be localized in particular areas of the crust. Data from the spacecraft's laser altimeter have provided the first three-dimensional views of the northern ice cap on Mars. Finally, new temperature data and close-up images of the Martian moon, Phobos, suggest that its surface consists of a powdery material at least 1 meter thick—most likely the result of millions of years of meteoroid impacts.

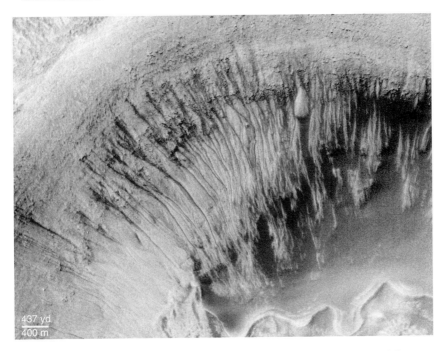

NASA's *Mars Global Surveyor* spacecraft took this high-resolution image of the north wall of a small crater located in the southwestern quarter of Newton Crater— a major surface feature on Mars. Scientists hypothesize that Newton Crater, a large basin about 178 miles (287 km) across, was probably formed by an asteroid impact more than three billion years ago. The small crater's north wall has many narrow gullies eroded into it. To some scientists, the presence of these gullies suggests that water and debris once flowed in ancient times on the surface of Mars. *(NASA/JPL/ Malin Space Science Systems)*

✧ Mars Odyssey 2001 Mission

NASA launched the *Mars Odyssey 2001* spacecraft to the Red Planet from Cape Canaveral Air Force Station on April 7, 2001. The robot spacecraft, previously called the *Mars Surveyor 2001 Orbiter,* was designed to determine the composition of the planet's surface, to detect water and shallow buried ice, and to study the ionizing radiation environment in the vicinity of Mars. The spacecraft arrived at the planet on October 24, 2001, successfully entered orbit, and then performed a series of aerobrake maneuvers to trim itself into the polar orbit around Mars for scientific data collection. The scientific mission began in January 2002.

Mars Odyssey has three primary science instruments: the thermal emission imaging system (THEMIS), a gamma-ray spectrometer (GRS), and the Mars radiation environment experiment (MARIE). THEMIS examined the surface distribution of minerals on Mars, especially those that can form only in the presence of water. The GRS determined the presence of 20 chemical elements on the surface of Mars, including shallow, subsurface pockets of hydrogen that act as a proxy for determining the amount and distribution of possible water ice on the planet. Finally, MARIE analyzed the Martian radiation environment in a preliminary effort to determine the potential hazard to future human explorers. The spacecraft collected scientific data until the end of its primary scientific mission on August 24, 2004. From that point, it began to function as an orbiting communications relay. It is currently supporting information transfer between the Mars Exploration Rover (MER) mission and scientists on Earth.

NASA selected the somewhat unusual name *Mars Odyssey 2001* for this important spacecraft as a tribute to the vision and spirit of space exploration that is embodied in the science-fact and science-fiction works of the famous British writer Sir Arthur C. Clarke.

✧ Mars Express Mission

The *Mars Express* spacecraft is part of a mission to Mars that was launched in June 2003 and developed by the European Space Agency (ESA) and the Italian Space Agency. After the 2,292-pound (1,042-kg) spacecraft arrived at the Red Planet in December 2003, its scientific instruments began to study the atmosphere and surface of Mars from a polar orbit. The main objective of *Mars Express* is to search from orbit for suspected subsurface water locations. The spacecraft also delivered a small lander spacecraft to investigate more closely the most suitable candidate site.

This small lander spacecraft was named *Beagle 2* in honor of the famous ship in which the British naturalist Charles Darwin (1809–82) made his great voyage of scientific discovery. After coming to rest on the surface of Mars, *Beagle 2* was to have performed exobiology and geochemistry research. The *Beagle 2* was scheduled to land on December 25, 2003. However, following deployment of the *Beagle 2* from its mother spacecraft, ESA ground controllers were unable to communicate with the probe and it was declared lost on February 6, 2004. Despite the problems with *Beagle 2*, the *Mars Express* spacecraft has functioned very well in orbit around the planet and accomplished its main mission of global high-resolution photogeology and mineralogical mapping of the Martian surface. In August 2004, the *Mars Express* also relayed images back to Earth from NASA's *Opportunity* (MER-B) surface rover as part of an international interplanetary networking demonstration.

In December 2006, scientists announced that data from the pioneering subsurface sounding radar altimeter (called MARSIS) on board ESA's

Mars Express orbiter indicated that the Red Planet has an older, craggier face buried beneath its smooth surface. This instrument has provided important new clues about the still mysterious geological history of Mars. MARSIS is the first subsurface radar used to explore a planet, and its investigative technique involves an evaluation of the echoes of transmitted radio waves that have penetrated below the planet's surface. The instrument's data strongly suggest that ancient impact craters lie beneath the smooth, low plains of Mars's northern hemisphere. MARSIS found evidence that these buried impact craters—ranging in diameter from about 81 miles (130 km) to 292 miles (470 km)—are present under much of the northern lowlands of Mars. In contrast to Earth, Mars displays a striking difference between its northern and southern hemispheres. Almost the entire southern hemisphere of the Red Planet has rough, heavily cratered highlands, while most of the planet's northern hemisphere is smoother and lower in elevation. The new finding by the *Mars Express* orbiter brings planetary scientists an important step closer to understanding one of the most interesting and enduring mysteries about the geological evolution and history of Mars.

✧ Mars Exploration Rover (MER) 2003 Mission

In the summer of 2003, NASA launched identical-twin Mars rovers that were to operate on the surface of the Red Planet during 2004. *Spirit* (MER-A) was launched by a Delta II rocket from Cape Canaveral on June 10, 2003, and successfully landed on Mars on January 4, 2004. *Opportunity* (MER-B) was launched from Cape Canaveral on July 7, 2003, by a Delta II rocket and successfully landed on the surface of Mars on January 25, 2004. Both landings resembled the successful airbag bounce-and-roll arrival demonstrated during the *Mars Pathfinder* mission.

Following arrival on the surface of the Red Planet, each rover drove off and began its surface exploration mission in a decidedly different location on Mars. *Spirit* (MER-A) landed in Gusev Crater, which is roughly 15 degrees south of the Martian equator. NASA mission planners selected Gusev Crater because it had the appearance of a crater lake bed. *Opportunity* (MER-B) landed at Terra Meridiani—a region of Mars that is also known as the Hematite Site because this location displayed evidence of coarse-grained hematite, an iron-rich mineral, which typically forms in water. Among this mission's principal scientific goals is the search for and characterization of a wide range of rocks and soils that hold clues to past water activity on Mars. By the end of 2006, both rovers continued to

NASA's Mars exploration rover (MER) on the surface of the Red Planet in 2004. *(NASA/JPL)*

function and move across Mars far beyond the primary mission goal of 90 days.

In early October 2006, NASA's long-lived robot rover *Opportunity* began to explore layered rocks in cliffs ringing the massive Victoria Crater on Mars. While *Opportunity* spent its first week at the crater, the *Mars Reconnaissance Orbiter (MRO)*—NASA's newest eye in the Martian sky—imaged the hardy rover and its surroundings from above. NASA mission controllers used the *MRO*'s high-resolution imagery to help guide *Opportunity* as it explored Victoria Crater. The efficient and timely blending of data from both robot systems is an excellent example of more-efficient planetary exploration through the integration of orbiter and lander/rover spacecraft. Exposed geological layers in the clifflike portions of the inner walls of Victoria Crater appear to record a longer span of Martian environmental history that the rover has studied in smaller craters.

With much greater mobility than the *Mars Pathfinder* minirover, each of these powerful new robot explorers has successfully traveled up to 330 feet (100 m) per Martian day across the surface of the planet. Each rover carries a complement of sophisticated instruments that allows it to search for evidence that liquid water was present on the surface of Mars in ancient times. *Spirit* and *Opportunity* have visited different regions of the planet. Immediately after landing, each rover performed reconnaissance of the particular landing site by taking panoramic (360°) visible (color) and infrared images. Then, using images and spectra taken daily by the rovers, NASA scientists at JPL used telecommunications and teleoperations to supervise the overall scientific program. With intermittent human guidance, the pair of mechanical explorers functioned like robot prospectors—examining particular rocks and soil targets and evaluating composition and texture at the microscopic level. Within two months after landing on Mars, *Opportunity* found geological evidence for an ancient Martian environment that was wet.

Each 407-pound (185-kg) rover has a set of five instruments with which to analyze rocks and soil samples. The instruments include a panoramic camera (Pancam), a miniature thermal emission spectrometer (Mini-TES), a Mössbauer spectrometer (MB), an alpha particle X-ray spectrometer (APXS), magnets, and a microscopic imager (MI). There is also a special rock abrasion tool (or RAT) that allowed each rover to expose fresh rock surfaces for additional study of interesting targets.

In January 2007, both *Spirit* and *Opportunity* celebrated the third anniversary of their successful landings on Mars. Although the robot rovers have begun to show some signs of wear and tear, they have also gotten a little smarter in the course of operating for more than three years on

This cylindrical-projection mosaic was assembled from images taken by the navigation camera on the Mars exploration rover *Spirit* on sol 107 (April 21, 2004) at a region dubbed "site 32" by mission control personnel at the Jet Propulsion Lab (JPL). *Spirit* is sitting east of Missoula Crater, no longer in the crater's ejecta field but on the outer plains. *(NASA/JPL)*

the surface of the Red Planet. Their unexpected longevity has provided NASA engineers a unique opportunity to field-test new capabilities. The extraterrestrial field tests not only helped improve the performance of the current rovers but also exerted influence on the design and performance of future rover systems. Specifically, both hardy rovers are testing four new skills that NASA engineers uploaded into their onboard computers.

One of the new capabilities enables each robot spacecraft to examine collected images and recognize certain features. Another new capability, visual target tracking, enables a moving rover to keep recognizing a designated landscape feature. With this capability, although the shape of a target rock may look different from a different angle, the robot rover now "knows" it is approaching the same (target) rock. The new software also improves the autonomy of each rover and helps it avoid potential hazards. Now, while wandering through potentially hazardous terrain, each rover can "think" several steps ahead.

As of the end of January 2007, *Spirit* and *Opportunity* have operated on the surface of Mars for nearly 12 times as long as their originally planned prime missions of 90 Martian days. *Spirit* has driven about 4.3 miles (6.9 km) and *Opportunity* about 6.1 miles (9.9 km).

Spirit has discovered evidence that water (in some form) has altered the mineral composition of certain soils and rocks in the older foothills above the plain where the rover landed. *Opportunity*'s key discovery since landing on Mars has been the collection of mineral and rock-texture evidence, which strongly suggests that water flowed over the surface and drenched at least one region of the Red Planet long ago.

✧ *Mars Reconnaissance Orbiter* (MRO)

NASA launched the *Mars Reconnaissance Orbiter (MRO)* on August 12, 2005, from Cape Canaveral Air Force Station using an Atlas III expendable booster. The primary mission of this spacecraft is to make high-resolution measurements of the planet's surface from orbit. It is equipped with a visible stereo imaging camera (HiRISE) with resolution much better than 3 feet (1 m) and a visible/near-infrared spectrometer (CRISM) to study the surface composition. Also on board the spacecraft is an infrared radiometer, an accelerometer, and the shallow subsurface sounding radar (SHARAD) provided by the Italian Space Agency to search for underground water. Tracking of the orbiter spacecraft will give scientists information on the gravity field of Mars.

The primary science objectives of the mission will be to look for evidence of past or present water, to study the weather and climate, and to identify landing sites for future missions. Data from *MRO* is allowing scientists to investigate complex terrain on Mars and to identify water-

Starting in 2006, NASA's *Mars Reconnaissance Orbiter* spacecraft began to take extremely high-resolution images of the planet's surface and to use its sounder to investigate scientifically interesting areas for the possibility of subsurface water. *(NASA/JPL)*

related landforms. The *MRO* is also assisting scientists as they search for sites on the Red Planet that exhibit stratigraphic or compositional evidence of water or hydrothermal activity. Instruments carried by the *MRO* can probe beneath the planet's surface for evidence of subsurface layering, water, and ice and can profile the internal structure of the Martian polar caps. Another important role that *MRO* is playing is to identify and characterize sites on Mars with the highest potential for future robot missions to the surface, including robot missions to collect soil and rock samples for analysis on Earth.

The *MRO* spacecraft has a height of 21 feet (6.5 m) and is topped by a 10-foot- (3-m-) diameter radio frequency antenna dish. The spacecraft has a width of 45 feet (13.6 m) from the tip of one extended solar panel to the tip of the other. The solar panels contain about 220 square feet (20 m²) of solar cells for electric power generation. At launch, *MRO* had a mass of

4,800 pounds (2,180 kg) with propellant for its 20 onboard thrusters making up about one-half of this mass.

In early September 2006, *MRO* completed the challenging task of shaping its orbit to a nearly circular, low altitude (about 160 mile [255 km]) polar orbit needed to scrutinize carefully the entire surface of the Red Planet. The spacecraft is now in its planned 24-month (two Earth years) science-mission phase. *MRO* can pour science data back to scientists on Earth more than 10 times faster than any previous Mars mission. Allowing the spacecraft to accomplish this great improvement in telemetry are a wider antenna dish, a faster onboard computer, and an amplifier powered by a larger solar-cell array. The information gathering will take place during one Martian year (roughly two Earth years), after which NASA will use the orbiter spacecraft as a communications relay satellite for future missions that land on the surface of Mars, such as the *Phoenix* lander spacecraft. Data from the *MRO* will also help NASA scientists select landing sites for future surface missions, such as the *Mars Science Laboratory*.

✧ Smarter Robots to the Red Planet

NASA's planned *Phoenix Mars Scout* will land in icy soils near the north polar permanent ice cap of the Red Planet and explore the history of water in these soils and any associated rocks. This sophisticated space robot serves as NASA's first exploration of a potential modern habitat on Mars and opens the door to a renewed search for carbon-bearing compounds, last attempted with the *Viking 1* and *2* lander spacecraft missions in the 1970s.

The *Phoenix* spacecraft is currently in development and will launch in August 2007. The robot explorer will land in May 2008 at a candidate site in the Martian polar region previously identified by the *Mars Odyssey* orbiter spacecraft as having high concentrations of ice just beneath the top layer of soil. *Phoenix* is a fixed-in-place lander, which means that it cannot move from one location to another on the surface of Mars. Rather, once the spacecraft has safely landed, it will stay there and use its robotic arm to dig the ice layer and bring samples to its suite of on-deck science instruments. These instruments will analyze samples directly on the Martian surface, sending science data back to Earth via radio signals, which will be collected by NASA's Deep Space Network.

The *Phoenix* spacecraft's stereo color camera and a weather station will study the surrounding environment, while its other instruments check excavated soil samples for water, organic chemicals, and conditions that could indicate whether the site was ever hospitable to life. Of special interest to exobiologists, the spacecraft's microscopes would reveal features as small as one one-thousandth the width of a human hair.

The *Phoenix* lander's science goals of learning about ice history and climate cycles on Mars complements the robot spacecraft's most exciting task—to evaluate whether an environment hospitable to microbial life may exist at the ice-soil boundary. One tantalizing question is whether cycles on Mars, either short-term or long-term, can produce conditions in which even small amounts of near surface water might stay melted. As studies of arctic environments on Earth have indicated, if water remains liquid only—even just for short periods during long intervals—life can persist, if other factors are right.

Building on the success of the two Mars Exploration Rover (MER) spacecraft, *Spirit* and *Opportunity*—which arrived on the surface of the Red Planet in January 2004—NASA's next mobile rover mission to Mars is being planned for arrival on the planet in late 2010. Called the *Mars Science Laboratory (MSL)*, this mobile robot will be twice as long and

This artist's rendering shows NASA's planned *Phoenix* robot lander spacecraft deployed on the surface of Mars (ca. 2008). The lander would use its robotic arm to dig into a spot in the water-ice rich northern polar region of Mars for clues concerning the Red Planet's history of water. The robot explorer would also search the landing site for environments that would be suitable for microscopic organisms (microbes). *(NASA)*

Sunset on Mars catches NASA's planned *Mars Science Laboratory (MSL)* in this artist's rendering. Once on the surface of the Red Planet (ca. 2010), the *Mars Science Laboratory* will analyze dozens of samples that will be scooped up from the soil and cored from rocks as the sophisticated robot spacecraft explores with greater range than any previous Mars rover. The MSL will investigate the past and present ability of Mars to support life. NASA engineers are considering the use of nuclear energy in the form of a radioisotope thermoelectric generator (RTG) to provide the robot an ample supply of electric power. This design option would give the robot rover a long operating lifespan under a variety of Martian environmental conditions. *(NASA/JPL)*

three times as massive as either *Spirit* or *Opportunity*. The *Mars Science Laboratory* will collect Martian soil samples and rock cores and analyze them on the spot for organic compounds and environmental conditions that could have supported microbial life in the past or possibly even now in the present.

After scientists have identified a geophysical signature for water from orbit around Mars or from surface rover mission investigations, the next important step is to drill in that location. The paramount goal of exobiologists is to "follow the water" in search of signs of life on Mars. Although scientists do not currently know much about subsurface conditions on the Red Planet, data from previous missions to Mars have provided abundant evidence that in ancient times the planet once had surface water, including streams and possibly shallow seas or even oceans. While there is photographic evidence for recent gullies, possibly cut by flowing water, there is no evidence for liquid water currently at the surface.

The icy, northern polar region of Mars is interesting to exobiologists because that is where the (frozen) water is—and where there is water (even in the form of ice), there could be life, existent or extinct. Drawing upon experience with some of the colder parts of Earth, as found, for example, in Iceland or Antarctica, scientists know that water can be stored as a mixture of frozen mud and ice in a layer of permafrost and beneath a permafrost layer, as liquid groundwater. So even if the ancient surface water on Mars evaporated, there may still be substantial reservoirs of water, in either liquid or frozen form, beneath the planet's surface.

To get to the zone where frozen water, and possibly dormant life might be present on Mars, scientists anticipate that they will have to drill or penetrate to a depth of about 660 feet (200 m). In all likelihood, any liquid water (if present) will be even deeper below the surface. Deep subsurface access on Mars presents unique engineering challenges. One approach to reach below the polar-region surface on Mars is to use a system called the cryobot ice-penetrating robot probe.

After a successful soft landing in the treacherous northern polar regions of Mars, the lander spacecraft would activate the ice-penetrating probe it carries. By heating the torpedo-shaped nose of the cryobot, the device is pulled by gravity down through the tunnel that the probe melts in the ice. Instruments carried within the body of the cryobot automatically take measurements and perform analyses of the gases and other materials encountered. As a smart robot, the cryobot probe will use innovative heating and steering features to maneuver around subsurface obstacles adroitly (primarily large rocks). Endowed with a high level of machine intelligence, the cryobot probe even has the ability to alter its downward course slightly to adjust for subsurface conditions or to exploit unexpected scientific opportunities, such as encountering a liquid-water aquifer deep beneath the frozen surface.

The cryobot moves through ice by melting the surface directly in front of it and allowing the liquid to flow around the robot probe and refreeze behind it. The probe takes measurements that characterize the encountered environment and then relays the scientific data up to the lander craft through a thin cable that is spooled out from its aft section as the robot probe descends into the frozen material. For ice layers of more than a mile (1.6 km) or so thick, it may be more practical to have the robot probe communicate back to the lander at the surface through a series of mini-radio-wave transceiver relays, which the probe deposits in the resolidified material as it descends.

The cryobot probe represents an innovative combination of active and passive melting systems. The cryobot method of subsurface penetration is

This is an artist's rendering of a cryobot—a robot probe for penetrating into the icy surface of a planet or a moon. The cryobot moves through ice by melting the surface directly in front of it while allowing liquid to flow around the torpedo-shaped robot probe and refreeze behind it. As it makes its molelike passage into an alien world, the cryobot's instruments take measurements of the encountered environment and send collected data back to the surface lander. On Mars, it appears that a communications cable could be used for penetration of shallower depths. On Europa, the thicker ice would require use of a network of miniradio-wave transceiver relays that are embedded in the ice. The use of semiautonomous steering and levels of artificial intelligence that promote fault management will help reduce the risk of the robot probe becoming trapped by subsurface obstructions, such as large rocks. (NASA/JPL)

more effective than conventional drilling techniques because it uses less power than mechanical cutting. Furthermore, since the cryobot travels downward in a self-sealing pathway through the ice, there is no deeply drilled hole that must be encased with massive steel tubes to prevent cave-in or collapse. Finally, the use of semiautonomous steering and fault management allow the probe to reduce the risk of becoming trapped by unanticipated subsurface conditions or obstructions. As mentioned in chapter 5, the cryobot would also be an ideal robot system for penetrating the ice crust of Europa.

✧ Mars Sample Return Mission (MSRM)

The purpose of a Mars Sample Return Mission (MSRM) is, as the name implies, to use a combination of robot spacecraft and lander systems to collect soil and rock samples from Mars and then return them to Earth for detailed laboratory analysis. A wide variety of options for this type of mission are being explored. For example, one or several small robot rover

This is an artist's rendering of a Mars Sample Return Mission (MSRM). The sample return spacecraft is shown departing the surface of the Red Planet after soil and rock samples that previously had been gathered by robot rovers were stored on board in a specially sealed capsule. To support planetary protection protocols, once in rendezvous orbit around Mars, the sample return spacecraft would use a mechanical device to transfer the sealed capsule of Martian soil samples to an orbiting Earth-return spacecraft ("mother ship") that would then take the samples back to Earth for detailed study by scientists. *(NASA/JPL; artist Pat Rawlings)*

vehicles could be carried and deployed by the lander vehicle. These rovers (under the control of operators on Earth) would travel away from the original landing site and collect a wider range of rock and soil samples for return to Earth.

Another option is to design a nonstationary, or mobile, lander that could travel (again guided by controllers on Earth) to various surface locations and collect interesting specimens. After the soil collection mission was completed, the upper portion of the lander vehicle would lift off from the Martian surface and rendezvous in orbit with a special "carrier" spacecraft. This automated rendezvous/return "carrier" spacecraft would remove the soil sample capsules from the ascent portion of the lander vehicle and then depart Mars orbit on a trajectory that would bring the samples back to Earth. After an interplanetary journey of about one year, this automated "carrier" spacecraft, with its precious cargo of Martian soil and rocks, would achieve orbit around Earth.

To avoid any potential problems of extraterrestrial contamination of Earth's biosphere by alien microorganisms that might possibly be contained in the Martian soil or rocks, the sample capsules might first be analyzed in a special human-tended orbiting quarantine facility. An alternate return mission scenario would be to bypass an Earth-orbiting quarantine process altogether and use a direct reentry vehicle operation to bring the encapsulated Martian soil samples to Earth.

Whichever sample return mission profile ultimately is selected, contemporary analysis of Martian meteorites (that have fallen to Earth) has stimulated a great scientific interest in obtaining well-documented and well-controlled "virgin" samples of Martian soil and rocks. Carefully analyzed in laboratories on Earth, these samples will provide a wealth of important and unique information about the Red Planet. These samples might even provide further clarification of the most intriguing question of all: Is there (or, at least, has there been) life on Mars? A successful Mars Sample Return Mission is also considered a significant and necessary step toward eventual human expeditions to Mars in this century.

✦ First Human–Crewed Expedition to Mars

The human-crewed expedition to Mars will most likely occur before the midpart of this century—possibly as early as 2035. Current expedition scenarios suggest a 600- to 1,000-day duration mission that most likely will start from Earth orbit, possibly powered by a nuclear thermal rocket. A total crew size of as many as 15 astronauts is anticipated. After hundreds of days of travel through interplanetary space, the first Martian explorers

This artist's rendering shows two astronaut–scientists (one an exobiologist and the other a geologist) exploring a promising sedimentary deposit on Mars ca. 2035. They are searching for fossils of ancient life on the Red Planet. *(NASA/JSC; artist, Pat Rawlings)*

will have about 30 days allocated for surface excursion activities on the Red Planet. Previous robot missions will have identified candidate sites worthy of direct detailed investigation by human explorers.

Searching for Life Elsewhere in the Solar System

In the not too distant past, it was quite popular of think of Venus as literally Earth's twin. People thought that since Venus's diameter, density, and gravity were only slightly less than Earth's, the cloud-enshrouded planet must be similar—especially since it had an obvious atmosphere and was a little nearer the Sun. Visions of a planet with oceans, tropical forests, giant reptiles, and possibly even primitive humans frequently appeared in science-fiction stories during the first half of the 20th century.

However, since the 1960s, visits by numerous American and Russian spacecraft dispelled all these pre–space age romantic fantasies. The data from these spacecraft missions clearly showed that the cloud-enshrouded planet was definitely not a prehistoric world that mirrored a younger Earth. Except for a few physical similarities of size and gravity, scientists now known that Earth and Venus are very different worlds. For example, the surface temperature on Venus approaches 932°F (500°C), its atmospheric pressure is more than 90 times that of Earth, it has no surface water, and its dense, hostile atmosphere with sulfuric acid clouds and an overabundance of carbon dioxide (about 96 percent) represents a runaway greenhouse of disastrous proportions.

Spacecraft missions throughout the solar system have resulted in a renaissance of interest and understanding about the formerly mysterious worlds (planets and moons) in this solar system. Celestial bodies once believed to have been life-bearing worlds have proven to be disappointingly barren, while other worlds now hold out the promise of sustaining perhaps primitive forms of extraterrestrial life. Besides Mars (see chapter 4), the leading candidate is the Jovian moon, Europa.

✧ Europa

For exobiologists, Jupiter's moon Europa remains one of the most intriguing places in the solar system. The moon has an icy shell that overlies a suspected deep-water (subsurface) ocean, and tidal flexing due to Jupiter's immense gravitational tug may provide energy for life. Until the space age, Europa remained little more than a dot of light. Flyby missions to Jupiter by the *Pioneer 10* and *11* spacecraft and the *Voyager 1* and *2* spacecraft in the 1970s and early 1980s tweaked a great deal of interest in this mysterious Galilean satellite. NASA's *Galileo* spacecraft provided even more tantalizing data.

The Galileo mission began on October 18, 1989, when the sophisticated spacecraft was carried into low Earth orbit by the space shuttle *Atlantis* and then started on its interplanetary journey by means of an inertial upper-stage (IUS) rocket. Relying on gravity-assist flybys to reach Jupiter, the *Galileo* spacecraft flew past Venus once and Earth twice. As it traveled through interplanetary space beyond Mars on its way to Jupiter, *Galileo* encountered the asteroids Gaspra (October 1991) and Ida (August 1993). *Galileo's* flyby of Gaspra on October 29, 1991, provided scientists their first-ever close-up look at a minor planet. On its final approach to Jupiter, *Galileo* observed the giant planet being bombarded by fragments of comet Shoemaker-Levy 9, which had broken apart. On July 12, 1995, the *Galileo* mother spacecraft separated from its hitchhiking companion (an atmospheric probe), and the two robot spacecraft flew in formation to their final destination.

SOME INTERESTING FACTS ABOUT EUROPA

Europa is the smooth, ice-covered moon of Jupiter, discovered in 1610 by the Italian scientist Galileo Galilei (1564–1642). Flyby visits by robot spacecraft lead scientists to think that this intriguing moon has a liquid-water ocean beneath its frozen surface. Europa has a diameter of 1,942 miles (3,124 km) and a mass of 10.6×10^{22} pounds (4.84×10^{22} kg). The moon is in synchronous orbit around Jupiter at a distance of 416,970 miles (670,900 km). An eccentricity of 0.009, an inclination of 0.47 degree, and a period of 3.551 (Earth) days further characterize the moon's orbit around Jupiter. The acceleration of gravity on the surface of Europa is 4.33 feet per second per second (1.32 m/s), and the icy moon has an average density of 188 pounds per cubic foot ($3,020 kg/m^3$). Next to Mars, exobiologists favor Europa as a leading candidate world within our solar system to possibly harbor some form of alien life.

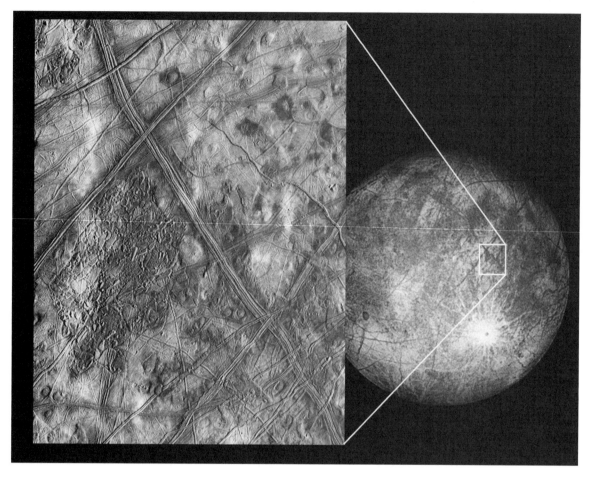

The image on the left shows a region of Europa's crust made up of blocks, which scientists think have broken apart and "rafted" into new positions. These features are the best geological evidence to date that Europa may have had a subsurface ocean at some time in its past. Combined with the geologic data, the presence of a magnetic field leads scientists to believe that a subsurface ocean is also probably present at Europa today. These images were obtained by NASA's *Galileo* spacecraft during September 7, 1996; December 1996; and February 1997 at a distance of 420,760 miles (677,000 km). *(NASA/JPL)*

On December 7, 1995, *Galileo* fired its main engine to enter orbit around Jupiter and gathered data that was transmitted from the atmospheric probe during that small robot's parachute-assisted descent into the Jovian atmosphere. During its two-year prime mission, the *Galileo* spacecraft performed 10 targeted flybys of Jupiter's major moons. In December 1997, the sophisticated robot spacecraft began an extended scientific mission that featured eight flybys of Jupiter's smooth, ice-covered moon Europa and two flybys of the pizza-colored, volcanic Jovian moon Io.

Galileo started a second extended scientific mission in early 2000. This second extended mission included flybys of the Galilean moons Io, Ganymede, and Callisto, plus coordinated observations of Jupiter with the *Cassini* spacecraft. In December 2000, *Cassini* flew past the giant planet to receive a much-needed gravity assist, which enabled the large spacecraft to eventually reach Saturn. *Galileo* conducted its final flyby of a Jovian moon in November 2002, when it zipped past the tiny inner moon Amalthea.

The encounter with Amalthea left *Galileo* on a course that would lead to an intentional impact into Jupiter in September 2003. NASA mission controllers deliberately crashed the *Galileo* mother spacecraft into Jupiter at the end of the space robot's very productive scientific mission to avoid any possibility of contaminating Europa with terrestrial microorganisms. As an uncontrolled derelict, the *Galileo* spacecraft might have eventually crashed into Europa sometime within the next few decades. Many exobiologists suspect that Europa has a life-bearing, liquid-water ocean underneath its icy surface. Since the *Galileo* spacecraft was probably harboring a variety of hitchhiking terrestrial microorganisms, scientists thought it prudent to avoid completely any possibility of contamination of Europa. The easiest way of resolving the potential problem was simply to dispose the retired *Galileo* spacecraft in the frigid, swirling clouds of Jupiter. So, NASA and JPL mission controllers accomplished this task, while they still maintained sufficient control over *Galileo's* behavior and trajectory.

Close-up study of Europa by NASA's *Galileo* spacecraft has provided scientists with tantalizing hints that this major moon of Jupiter might possess a liquid-water ocean beneath its icy crust. Where there is liquid water, there is the intriguing possibility that alien life forms (ALFs) may be found. So NASA strategic planners are examining (on a conceptual basis) several future robot missions that would to quite literally "break the ice" in the search for alien life beyond Earth.

NASA's proposed Jupiter Icy-Moons Orbiter (JIMO) mission involves an advanced technology robot spacecraft that would orbit three of Jupiter's most intriguing moons—Callisto, Ganymede, and Europa. All three planet-sized moons may have liquid-water oceans beneath their icy surfaces. Following up on the historic discoveries made by the *Galileo* spacecraft, the JIMO mission would make detailed studies of the composition, geologic history, and potential for sustaining life on each of these three large icy moons. The mission's proposed science goals include: to scout for potential life on the moons; to investigate the origin and evolution of the moons; to explore the radiation environment around each moon; and to determine how frequently each moon is battered by space debris.

The JIMO spacecraft would pioneer the use of electric propulsion that is powered by a nuclear-fission reactor. Contemporary electric propulsion technology—successfully tested on the NASA's *Deep Space 1* spacecraft—

would allow the planned JIMO spacecraft to orbit three different moons during a single mission. Current spacecraft, like *Cassini*, have enough onboard propulsive thrust capability (on arrival at a target planet) to orbit that single planet and then use various orbits to fly by any moons or other objects of interest, such as ring systems. In contrast, the JIMO's proposed nuclear electric propulsion system would have the necessary long-term thrust capability to gently maneuver through the Jovian system and allow the spacecraft to orbit each of the three icy moons of interest successfully.

This is an artist's rendering of NASA's proposed Project Prometheus nuclear-reactor powered, ion-propelled spacecraft that will enter the Jovian system, ca. 2015. The Jupiter Icy-Moons Orbiter (JIMO) mission would perform detailed scientific studies of Callisto, Ganymede, and Europa (in that order), searching for liquid-water oceans beneath their frozen surfaces. Europa is of special interest to the scientific community because its suspected ocean of liquid water could contain alien life-forms. *(NASA/JPL)*

Another very interesting conceptual robot space mission to search for life on Europa involves an orbiting mother spacecraft, a cryobot, and a hydrobot. The *Europa Orbiter* would serve as the robot command post for the entire mission. Once in orbit around Europa, this robot spacecraft would release a lander robot to a special location on the surface, identified as being of great scientific interest to exobiologists and other investigators. After it soft-lands on the ice-covered surface of Europa, the lander deploys a large cryobot probe, which also contains a hydrobot. The cryobot probe melts its way down through the icy crust of Europa until it reaches the suspected subsurface ocean. The cryobot has left a trail of radio transponders behind in the resolidified ice to communicate with the lander, which in turn relays data to the orbiting "mother spacecraft" in burst transmission mode. The mother spacecraft (*Europa Orbiter*) would keep scientists back on Earth informed of the mission's progress, but this is a totally automated operation without any direct human supervision of the orbiter spacecraft, the cryobot, or the hydrobot.

Once the cryobot has penetrated through Europa's thick icy crust and found the currently suspected subsurface ocean, the torpedo-shaped robot probe releases an autonomous, self-propelled underwater robot, called the hydrobot. The hydrobot scoots off and starts to take scientific measurements of its aquatic environment. The robot submarine also diligently investigates the waters of Europa for signs of alien life. Data from the hydrobot are relayed back to Earth via the cryobot, the surface lander, and the orbiting mother spacecraft. The following figure is an artist's rendering of what would be one of the great scientific discoveries of this century. This *hypothesized* scene shows the hydrobot examining an underwater thermal vent and various alien life-forms (ALFs) in Europa's subsurface ocean. The team of very smart robots (orbiter, lander, cryobot, and hydrobot) in this postulated scenario has allowed their human creators to discover life on another world in the solar system. Whether or not life actually exists on Europa, this type of advanced space exploration mission with a team of future robots exercising collective machine intelligence will be remarkable and will serve as a precursor to sets for even more exciting missions. As discussed in chapter 6, aerospace engineers would have to exercise a great deal of care in preventing the forward contamination of Europa by any future spacecraft that was sent from Earth to investigate this intriguing moon.

✧ Titan

With its rich atmosphere of organic molecules, for many exobiologists, Titan, the largest moon of the planet Saturn, is a dream natural laboratory. Titan's atmosphere is composed of nitrogen (N_2) and methane

This artist's rendering shows a proposed ice-penetrating cryobot (background) and a submersible hydrobot (foreground)—an intriguing advance robot combination that could be used to explore the suspected ice-covered ocean on Jupiter's moon Europa. In this scenario, a lander robot would arrive on Europa's surface and deploy the cryobot/hydrobot package, remaining on the surface to function as a communications relay station. The cryobot would melt its way through the ice cover and then deploy the hydrobot into the ice-covered ocean. The hydrobot is a self-propelled underwater vehicle that could analyze the chemical composition of the subsurface ocean and search for signs of alien life. The artwork here shows the autonomous robot submarine (hydrobot) examining a *hypothesized* underwater thermal vent and various alien aquatic life-forms gathered around this life-sustaining phenomenon. *(NASA/JPL)*

(CH_4) gas. Ultraviolet radiation from the Sun can break up these molecules, leading to the formation of complex organic molecules. In the past, exobiologists often suggested that the composition of Titan's atmosphere closely resembled that of an early Earth, before life began on humans' home planet. One of the most pressing questions that circulated within the scientific community concerning Titan involved the issue of how complex the organic molecules were in Titan's opaque, smoggy atmosphere. A companion issue raised by some exobiologists involved the possibility of cold-loving extremophiles arising on Titan as a result of the frigid organic molecule soup. The *Cassini/Huygens* spacecraft mission to

Saturn provided some answers and also raised some new, intriguing questions about this large, cloud-enshrouded moon of Saturn with a dense nitrogen-rich atmosphere.

SOME INTERESTING FACTS ABOUT TITAN

Titan is the largest Saturnian moon; approximately 3,200 miles (5,150 km) in diameter, it was discovered in 1655 by the Dutch astronomer Christiaan Huygens (1629–95). It has a sidereal period of 15.945 days and orbits at a mean distance of 759,385 miles (1,221,850 km) from the planet. Titan is the second-largest moon in the solar system and the only one that is known to have a dense atmosphere. The atmospheric chemistry presently taking place on Titan may be similar to those processes that occurred in Earth's atmosphere several billion years ago.

Larger in size than the planet Mercury, Titan has a density that appears to be about twice that of water ice. Scientists believe, therefore, that it may be composed of nearly equal amounts of rock and ice. Titan's surface is hidden from the normal view of spacecraft cameras by a dense, optically thick photochemical haze whose main layer is about 186 miles (300 km) above the moon's surface. Measurements prior to the July 2004 arrival of the *Cassini* spacecraft (with its hitchhiking companion *Huygens* probe) indicated that Titan's atmosphere is mostly nitrogen. The existence of carbon-nitrogen compounds on Titan is possible because of the great abundance of both nitrogen and hydrocarbons.

To date, the Cassini/Huygens mission has revealed that Titan is an extraordinary world that resembles Earth in many respects, especially meteorology, geomorphology, and fluvial activity—but with different ingredients. The images (collected by radar) show strong evidence for surface erosion due to liquid flows, possibly methane. The *Huygens* probe in-situ sampling data confirmed the presence of complex organic chemistry, which reinforces the notion that Titan is a promising place to observe the molecules that may have been the precursors of the building blocks of life on Earth.

In June 2005, the *Cassini* spacecraft took a set of radar images of the surface of Titan that revealed a unique, dark, lakelike feature with smooth shorelike boundaries. Some scientists, after examining the imagery, suggested that this intriguing dark feature in Titan's south polar region may be the site of a past or present lake of liquid hydrocarbons. The suspected lake area measures 145 miles (233 km) by 45 miles (73 km), which is about the size of Lake Ontario located on the U.S.-Canadian border. The interesting surface feature lies in Titan's cloudiest region—an area of the cloud-enshrouded moon that scientists suspect to be also the most likely location of recent methane rainfall.

Other scientists have offered a different interpretation of the dark surface feature. They suggest that this surface feature was once a lake but has since dried up, leaving behind dark deposits of hydrocarbons. A third interpretation is that this distinctive feature is simply a broad depression, which has (over time) filled with dark, solid hydrocarbons that have precipitated from Titan's atmosphere onto the moon's surface. Whatever this lakelike feature turns out to be, its presence is just one of the many new puzzles that Titan is tossing at scientists as the *Cassini* spacecraft continues to fly by and image this intriguing moon.

The *Cassini/Huygens* spacecraft mission was successfully launched by a mighty Titan IV–Centaur vehicle on October 15, 1997, from Cape Canaveral Air Force Station, Florida. This mission is a joint NASA and European Space Agency (ESA) project to conduct detailed exploration of Saturn, its major moon Titan, and its complex system of other moons. Following the example of the *Galileo* spacecraft, the *Cassini* spacecraft also took a gravity-assisted tour of solar system. The spacecraft eventually reached Saturn following a Venus-Venus-Earth-Jupiter gravity-assist (VVEJGA) trajectory. After a nearly seven-year journey through interplanetary space covering 2.2 billion miles (3.5 billion km), the *Cassini* spacecraft arrived at Saturn on July 1, 2004 (Eastern daylight time).

The very large and complex robot spacecraft is named in honor of the Italian-born French astronomer Giovanni Domenico Cassini (1625–1712), who was the first director of the Royal Observatory in Paris and conducted extensive observations of Saturn. The *Huygens* probe is named in honor of the Dutch astronomer Christiaan Huygens (1629–95), who discovered Titan in 1655.

The most critical phase of the mission after launch was Saturn orbit insertion (SOI). When *Cassini* arrived at Saturn, the sophisticated robot spacecraft fired its main engine for 96 minutes to reduce its speed and to allow it to be captured as a satellite of Saturn. Passing through a gap between Saturn's F and G rings, the intrepid spacecraft successfully swung close to the planet and began the first of some six-dozen orbits that it will complete during its four-year primary mission.

The arrival period provided a unique opportunity to observe Saturn's rings and the planet itself since this was the closest approach the spacecraft will make to Saturn during the entire mission. As anticipated, the *Cassini* spacecraft went right to work on arrival and provided scientific results.

Scientists examining Saturn's contorted F ring, which has baffled them since its discovery, have found one small body, possibly two, orbiting in the F ring region, and a ring of material associated with Saturn's moon Atlas. *Cassini*'s close-up look at Saturn's rings revealed a small object moving near the outside edge of the F ring, interior to the orbit of Saturn's moon Pandora. This tiny object, which is about 3.1 miles (5 km) in diameter, has been provisionally assigned the name S/2004 S3. It may be a tiny moon that orbits Saturn at a distance of 87,600 miles (141,000 km) from Saturn's center. This object is located about 620 miles (1,000 km) from Saturn's F ring. A second object, provisionally called S/2004 S4, has also been observed in the initial imagery provided by the *Cassini* spacecraft. About the same size as S/2004 S3, this object appears to exhibit some strange dynamics, which take it across the F ring.

In the process of examining the F-ring region, scientists also detected a previously unknown ring, now called S/2004 1R. This new ring is associated with Saturn's moon Atlas. The ring is located 85,770 miles (138,000 km) from the center of Saturn in the orbit of the moon Atlas, between the A ring and the F ring. Scientists estimate the ring has a width of 185 miles (300 km).

On arrival at Saturn and the successful orbit insert burn (July 2004), the *Cassini* spacecraft began its extended tour of the Saturn system. This orbital tour involves at least 76 orbits around Saturn, including 52 close encounters with seven of Saturn's known moons. The *Cassini* spacecraft's orbits around Saturn are being shaped by gravity-assist flybys of Titan. Close flybys of Titan also permit high-resolution mapping of the intriguing, cloud-shrouded moon's surface. The *Cassini* orbiter spacecraft carries an instrument called the Titan imaging radar, which can see through the opaque haze covering that moon to produce vivid topographic maps of the surface.

The size of these orbits, their orientation relative to Saturn and the Sun, and their inclination to Saturn's equator are dictated by various scientific requirements. These scientific requirements include: imaging radar coverage of Titan's surface; flybys of selected icy moons, Saturn, or Titan; occultations of Saturn's rings; and crossings of the ring plane.

The *Cassini* orbiter will make at least six close, targeted flybys of selected icy moons of greatest scientific interest—namely, Iapetus, Enceladus, Dione, and Rhea. Images taken with *Cassini's* high-resolution telescopic cameras during these flybys will show surface features equivalent in spatial resolution to the size of a major league baseball diamond. At least two dozen more distant flybys (at altitudes of up to 62,000 miles [100,000 km]) will also be made of the major moons of Saturn—other than Titan. The varying inclination of the *Cassini* spacecraft's orbits around Saturn will allow the spacecraft to conduct studies of the planet's polar regions, as well as its equatorial zone.

In addition to the *Huygens* probe, Titan is now undergoing scientific investigation by the *Cassini* orbiter. The spacecraft is executing 45 targeted, close flybys of Titan, Saturn's largest moon—some flybys as close as about 590 miles (950 km) above the surface. Titan is the only Saturn moon large enough to enable significant gravity-assist changes in *Cassini's* orbit. Accurate navigation and targeting of the point at which the *Cassini* orbiter flies by Titan will be used to shape the orbital tour. This mission planning approach is similar to the way the *Galileo* spacecraft used its encounters of Jupiter's large moons (the Galilean satellites) to shape its very successful scientific tour of the Jovian system.

As currently planned, the prime mission tour of the *Cassini* spacecraft will end on June 30, 2008. This date is four years after arrival at Saturn and 33 days after the last Titan flyby, which will occur on May 28, 2008. The aim point of the final flyby is being chosen (in advance) to position *Cassini* for an additional Titan flyby on July 31, 2008—providing mission controllers with the opportunity to proceed with more flybys during an extended mission, if resources (such as the supply of attitude-control propellant) allow. Nothing in the present design of the orbital tour of the Saturn system now precludes an extended mission.

The *Cassini* spacecraft, which originally included the orbiter and the *Huygens* probe, is the largest and most complex interplanetary spacecraft ever built. The orbiter spacecraft alone has a dry mass of 4,675 pounds (2,125 kg). When the 704 pound (320 kg) *Huygens* probe and a launch vehicle adapter were attached and 6,890 pounds (3,130 kg) of attitude-control and maneuvering propellants were loaded, the assembled spacecraft acquired a total launch mass of 12,570 pounds (5,712 kg). At launch, the fully assembled *Cassini* spacecraft stood 22 feet (6.7 m) high and 12.9 feet (4 m) wide.

The Cassini mission involves a total of 18 science instruments, six of which are contained in the wok-shaped *Huygens* probe. This ESA-sponsored probe was detached from the main orbiter spacecraft after *Cassini* on December 25, 2004, and successfully conducted its own scientific investigations as it plunged into the atmosphere of Titan on January 14, 2005. The probe's science instruments included: the aerosol collector pyrolyzer, descent imager and spectral radiometer, Doppler wind experiment, gas chromatograph and mass spectrometer, atmospheric structure instrument, and surface science package.

The *Cassini* spacecraft's science instruments include: a composite infrared spectrometer, imaging system, ultraviolet imaging spectrograph, visual and infrared mapping spectrometer, imaging radar, radio science, plasma spectrometer, cosmic-dust analyzer, ion and neutral mass spectrometer, magnetometer, magnetospheric imaging instrument, and radio and plasma wave science. Telemetry from the spacecraft's communications antenna is also being used to make observations of the atmospheres of Titan and Saturn and to measure the gravity fields of the planet and its satellites.

The Cassini mission (including *Huygens* probe and orbiter spacecraft) is designed to perform a detailed scientific study of Saturn, its rings, its magnetosphere, its icy satellites, and its major moon, Titan. The *Cassini* orbiter's scientific investigation of the planet Saturn includes: cloud properties and atmospheric composition; winds and temperatures; internal structure and rotation; the characteristics of the ionosphere; and the origin and evolution of the planet. Scientific investigation of the Saturn ring system includes: structure and composition; dynamic processes within the rings; the interrelation of rings and satellites; and the dust and micrometeoroid environment.

Saturn's magnetosphere involves the enormous magnetic bubble surrounding the planet that is generated by its internal magnet. The

HUYGENS SPACECRAFT

The *Huygens* probe was carried to the Saturn system by the *Cassini* orbiter spacecraft. Bolted to the *Cassini* mother spacecraft and fed electrical power through an umbilical cable, *Huygens* rode along during the nearly seven-year journey largely in a sleep mode. However, mission controllers did awaken the robot probe about every six months for three-hour-duration instrument and engineering checkups. *Huygens* is sponsored by the European Space Agency and named after the Dutch physicist and astronomer Christiaan Huygens (1629–95), who first described the nature of Saturn's rings and discovered its major moon, Titan, in 1655.

The *Cassini* spacecraft's second Titan flyby on December 13, 2004, left the spacecraft (which was still carrying the hitchhiking *Huygens* probe) on a trajectory that, if uncorrected, would lead to a subsequent flyby of Titan at an altitude of about 2,860 miles (4,600 km). To get the *Huygens* probe traveling into Titan's atmosphere at just the right angle, the *Cassini* mother spacecraft performed a targeting maneuver before it released its hitchhiking robot companion. On December 17, the Cassini spacecraft completed a precise targeting maneuver that shaped its course and pointed the cojoined robot spacecraft team on a direct impact trajectory to Titan.

On December 25, 2005 (at 02:00 universal time coordinated), the spin/eject device separated *Huygens* from *Cassini* with a relative speed of 1.1 feet per second (0.35 m/s) and a spin rate of 7.5 revolutions per minute. As a result of these successful maneuvers and actions, the spin-stabilized atmospheric probe was targeted for a southern-latitude landing site on the dayside of Titan. To support a variety of mission needs and parameters, the probe entry angle into Titan's atmosphere was set at a relatively steep 65 degrees. ESA mission controllers selected this entry angle to give the probe the best opportunity to reach the surface of Titan. Following probe separation, the *Cassini* orbiter performed some final maneuvers to avoid crashing into Titan and to position itself to collect data from *Huygens* as it descended into Titan's opaque, nitrogen-rich atmosphere.

On January 14, 2005, after coasting for 20 days, the *Huygens* probe reached the desired entry altitude of 790 miles (1,270 km) above Titan and started its parachute-assisted descent into the moon's atmosphere. Within five minutes, the probe began to transmit its science data to the *Cassini* orbiter as it floated down through Titan's atmosphere.

During the first part of the probe's atmospheric plunge, instruments on board *Huygens* were controlled by a timer. For the final 6 to 12 miles (10 to 20 km) of descent, a radar altimeter on board the probe controlled its scientific instruments on the basis of altitude. About 138 minutes after starting its plunge into Titan's upper atmosphere, the *Huygens* probe came to rest on the moon's surface. Images of the site were collected just before landing. The probe survived impact on a squishy surface that was neither liquid nor frozen solid—the two candidate surface conditions most frequently postulated by planetary scientists for this cloud-enshrouded moon. As the *Cassini* orbiter disappeared over the horizon, the mother spacecraft stopped collecting data from its hardworking robot companion. The probe had been continuously transmitting data for about four and one-half hours.

magnetosphere also consists of the electrically charged and neutral particles within this magnetic bubble. Scientific investigation of Saturn's magnetosphere includes: its current configuration; particle composition, sources and sinks; dynamic processes; its interaction with the solar wind, satellites, and rings; and Titan's interaction with both the magnetosphere and the solar wind.

During the orbit tour phase of the mission (from July 1, 2004, to June 30, 2008), the *Cassini* orbiter spacecraft will perform many flyby encounters of all the known icy moons of Saturn. As a result of these numerous satellite flybys, the spacecraft's instruments will investigate: the characteristics and geologic histories of the icy satellites; the mechanisms for surface modification; surface composition and distribution; bulk composition and internal structure; and interaction of the satellites with Saturn's magnetosphere.

The moons of Saturn are diverse—ranging from the planetlike Titan to tiny, irregular objects only tens of miles (km) in diameter. Scientists currently believe that all of these bodies (except for perhaps Phoebe) hold not only water ice but also other chemical components, such as methane, ammonia, and carbon dioxide. Before the advent of robotic spacecraft in space exploration, scientists believed that the moons of the outer planets were relatively uninteresting and geologically dead. They assumed that (planetary) heat sources were not sufficient to have melted the mantles of these moons enough to provide a source of liquid, or even semiliquid, ice, or silicate slurries.

As if Titan was not providing scientists with enough surprises, in March 2006, the *Cassini* spacecraft appears to have found evidence of liquid-water reservoirs that erupt in Yellowstone Park–like geysers on Saturn's moon Enceladus. The rare occurrence of liquid water so near the surface raises many new scientific questions about this mysterious moon of Saturn. One current speculation is that the jets of water being ejected from Enceladus might be erupting from near-surface pockets of liquid water above 32°F (0°C) like cold versions of Yellowstone Park's Old Faithful geyser on Earth. This discovery placed Enceladus in an exclusive solar-system club—that of planetary bodies where active volcanism exists. Jupiter's moon Io, Earth, and possibly Neptune's moon Triton are the other currently known members of that exclusive group. The presence of liquid water, perhaps just 100 feet (30 m) or less below the moon's icy crust now makes Enceladus one of the most exciting places in the solar system.

Scientists are quick to point out that this recent discovery has given the search for liquid water beyond Earth a dramatic new turn. The type of evidence for liquid water on Enceladus is very different from what scientists have seen at Jupiter's moon Europa. On Europa, the evidence from surface geological features suggests an internal liquid-water ocean. On Enceladus,

SOME INTERESTING FACTS ABOUT ENCELADUS

Enceladus is a moon of Saturn that was discovered in 1789 by the British astronomer Sir William Herschel (1738–1822). This relatively small moon has a diameter of approximately 310 miles (500 km) and orbits around Saturn at a distance of 147,898 miles (238,020 km) with a period of rotation of 1.37 days. The surface of Enceladus is dominated by fresh, clean ice, giving the moon an albedo of almost one. Because Enceladus reflects almost 100 percent of the incident sunlight, the moon has a surface temperature of only -330°F (-201°C). Since early 2005, the *Cassini* spacecraft has uncovered many mysteries about this intriguing moon. The *Cassini* spacecraft's magnetometer discovered an atmosphere around Enceladus, providing scientists evidence that gases may be escaping from the moon's surface or interior. In November 2005, the *Cassini* spacecraft's visual and infrared mapping spectrometer measured the spectrum of plumes originating from the south pole of the icy moon, providing scientists a very clear signature of small ice particles. Images of plumes of icy materials streaming from Enceladus's south pole, suggest that the moon has Yellowstone Park–like geysers fed by near-surface reservoirs of liquid water.

the evidence is direct observation (by instruments on board the *Cassini* spacecraft) of water vapor venting from liquid-water sources close to the surface. Scientists will be given another close-up look at Enceladus in the spring of 2008 when the *Cassini* spacecraft is scheduled to fly past this intriguing moon within 220 miles (350 km) of its surface.

✦ Exploring Small Bodies in the Solar System

Just two decades ago, scientists did not have very much specific information about the small bodies in the solar system, such as comets and asteroids. There was a great deal of speculation about the true nature of a comet's nucleus, and no one had ever seen the surface of an asteroid up close. All that changed very quickly when robot spacecraft missions flew past, imaged, sampled, probed, and even landed on several of these interesting celestial objects. This section of the chapter discusses NASA's *Stardust* spacecraft and its comet sampling missions, which represents one of the most significant small-body missions that have taken place.

Asteroids and comets are believed to be the ancient remnants of the earliest years of the formation of the solar system, which took place more than four billion years ago. From the beginning of life on Earth to the spectacular collision of comet Shoemaker-Levy 9 with Jupiter (in July 1994), these so-called small bodies influence many of the fundamental processes that have shaped the planetary neighborhood in which Earth resides.

Scientists currently believe that asteroids are the primordial material that was prevented by Jupiter's strong gravity from accreting (accumulating) into a planet-size body when the solar system was born about 4.6 billion years ago. It is estimated that the total mass of all the asteroids (if assembled together) would comprise a celestial body about 932 miles (1,500 km) in diameter—an object less than half the size (diameter) of the Moon.

NASA's *Galileo* spacecraft was the first to observe an asteroid close up, flying past main-belt asteroids Gaspra and Ida in 1991 and 1993, respectively. Gaspra and Ida proved to be irregularly shaped objects, rather like potatoes, riddled with craters and fractures. The *Galileo* spacecraft also discovered that Ida had its own moon—a tiny body called Dactyl in orbit around its parent asteroid. Astronomers suggest that Dactyl may be a fragment from past collisions in the asteroid belt.

A comet is a dirty ice rock consisting of dust, frozen water, and gases that orbits the Sun. As a comet approaches the inner solar system from deep space, solar radiation causes its frozen materials to vaporize (sublime), creating a coma and a long tail of dust and ions. Scientists think that these icy planetesimals are the remainders of the primordial material from which the outer planets were formed billions of years ago. As confirmed by spacecraft missions, a comet's nucleus is a type of dirty ice ball, consisting of frozen gases and dust. While the accompanying coma and tail may be very large, comet nuclei generally have diameters of only a few tens of miles (kilometers) or less.

The primary objective of NASA's Discovery class *Stardust* mission was to fly by the comet P/Wild 2 and collect samples of dust and volatiles in the coma of this comet. NASA launched the *Stardust* spacecraft from Cape Canaveral Air Force Station, Florida, on February 7, 1999, using an expendable Delta II rocket. Following launch, the spacecraft successfully achieved an elliptical, heliocentric orbit. By midsummer 2003, it had completed its second orbit of the Sun. The spacecraft then successfully flew by the nucleus of comet Wild 2 on January 2, 2004. When *Stardust* flew past the comet's nucleus, it did so at an approximate relative velocity of 6.1 kilometers per second. At closest approach during this close encounter, the spacecraft came within 155 miles (250 km) of the comet's nucleus and returned images of the nucleus. The spacecraft's dust-monitor data indicated that many particle samples were collected. *Stardust* then traveled on a trajectory that brought it near Earth in early 2006. The comet material samples that were collected had been stowed and sealed in the special sample storage vault of the reentry capsule carried on board

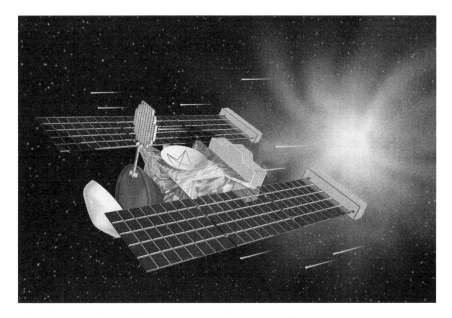

This artist's rendering shows NASA's *Stardust* spacecraft encountering comet Wild 2 (in January 2004) and collecting dust and volatile material samples in the coma of the comet. The robot spacecraft collected, stowed, and sealed the comet material samples in the special storage vault of an Earth–return reentry capsule, which was also carried on board *Stardust*. On January 15, 2006, the sample capsule successfully returned to Earth. *(NASA/JPL)*

the *Stardust* spacecraft. As the spacecraft flew past Earth in mid-January 2006, it ejected the sample capsule. The sample capsule descended through Earth's atmosphere and was recovered successfully in the Utah desert on January 15, 2006.

In inspecting the samples of comet material obtained by the *Stardust* spacecraft, scientists were able to confirm some anticipated results but were also surprised. For example, the returned samples show high-temperature materials (like olivine, the green Hawaiian beach sand) from the coolest parts of the solar system. It now appears that comets are really a mixture of materials formed at all temperatures, characteristic of places near the Sun and very far away from it. The olivine components included iron, magnesium, and other elements. The samples from comet Wild 2 included other high-temperature materials that contain calcium, aluminum, and titanium, as well as the silicate mineral forsterite, which can be found on Earth in gemstones called peridot. The preliminary analyses performed by approximately 200 scientists from around the world resulted in a general consensus that many of the comet particles are constructed like loose dirt clods, that is, composed of both large strong rocks as well as fine powdery materials. One of the most exciting preliminary conclusions is the suggestion that the comet is a mix of both stardust grains from other stars as well as materials formed in the solar system. This hypothesis would explain the puzzle of how the collected comet samples could contain some of the "hottest materials found in the coldest places." If this preliminary interpretation of the data is correct, then NASA's selection of the name *Stardust* for both the spacecraft and mission would have proven very appropriate.

Extraterrestrial Contamination: War of the Micro-biological Worlds

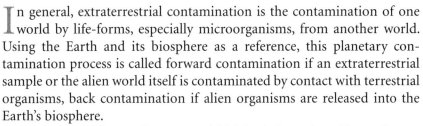

In general, extraterrestrial contamination is the contamination of one world by life-forms, especially microorganisms, from another world. Using the Earth and its biosphere as a reference, this planetary contamination process is called forward contamination if an extraterrestrial sample or the alien world itself is contaminated by contact with terrestrial organisms, back contamination if alien organisms are released into the Earth's biosphere.

Here on Earth, an alien terrestrial biological species will usually not survive when introduced into a new ecological system because it is unable to compete with native species that are better adapted to the environment. Once in a while, however, alien species actually thrive because the new environment is very suitable and because indigenous life-forms are unable to defend themselves successfully against these alien invaders. When this "war of biological worlds" occurs, the result might very well be a permanent disruption of the host ecosphere, with severe biological, environmental, and possibly economic consequences.

Of course, the introduction of an alien species into an ecosystem is not always undesirable. Many European and Asian vegetables and fruits, for example, have been introduced successfully and profitably into the North American environment. However, any time a new organism is released in an existing terrestrial ecosystem, a finite amount of risk is also introduced.

Frequently, alien organisms that destroy resident species are microbiological life-forms. Such microorganisms may have been nonfatal in their native habitat, but once released in the new ecosystem, they become unrelenting killers of native life-forms that are not resistant to them. In past centuries on Earth, entire human societies fell victim to alien organisms against which they were defenseless, as were, for example, the rapid spread of diseases that were transmitted to native Polynesians and American Indians by European explorers.

But an alien organism does not have to infect humans directly to be devastating. Who can ignore easily the consequences of the potato blight fungus that swept through Europe and the British Isles in the 19th century, causing a million people to starve to death in Ireland alone?

In the space age, it is obviously of extreme importance to recognize the potential hazard of extraterrestrial contamination (forward or back). Before any species is introduced intentionally into another planet's environment, scientists must determine carefully not only whether the organism is pathogenic (disease causing) to any indigenous species but also whether the new organism will be able to force out native species—with destructive impact on the original ecosystem. The introduction of rabbits into the Australian continent is a classic terrestrial example of a nonpathogenic life-form creating immense problems when introduced into a new ecosystem. The rabbit population in Australia simply exploded in size because of their high reproduction rate, which was essentially unchecked by native predators.

✧ Quarantine Protocols

At the start of the space age, scientists were already aware of the potential extraterrestrial-contamination problem—in either direction. Quarantine protocols (procedures) were established to avoid the forward contamination of alien worlds by outbound unmanned spacecraft, as well as the back contamination of the terrestrial biosphere when lunar samples were returned to Earth as part of the Apollo program. For example, the United States is a signatory to the 1967 Treaty on Principles Governing the Activities of States in the Exploration and Use of Outer Space, including the Moon and Other Celestial Bodies. This important international agreement establishes the legal requirements by which a signatory nation must take technical and operational measures to avoid forward and back contamination of planetary bodies during space exploration. The Committee on Space Research (COSPAR) of the International Council of Scientific Unions monitors a signatory nation's planetary contamination-prevention efforts.

A quarantine is basically a forced isolation to prevent the movement or spread of a contagious disease. Historically, quarantine was the period during which ships suspected of carrying persons or cargo (for example, produce or livestock) infected with contagious diseases were detained at their port of arrival. The length of the quarantine, generally 40 days, was considered sufficient to cover the incubation period of most highly infectious terrestrial diseases. If no symptoms appeared at the end of the quarantine, then the travelers were permitted to disembark, or the cargo was permitted to unload. In modern times, the term *quarantine* has obtained a new meaning, namely, that of holding a suspect organism or infected person in strict isolation until it is no longer capable of transmitting the disease. With the Apollo Project and the advent of the lunar quarantine, the term now has elements of both meanings. Of special interest in future space missions to the planets and their major moons is how scientists and aerospace engineers avoid the potential hazard of back contamination of Earth's environment when robot spacecraft and human explorers bring back samples of alien worlds for more detailed examination in laboratories on Earth.

Because the planet Mars and the Jovian moon Europa are potential life-bearing alien worlds, all space exploration activities involving these two celestial bodies merit very special attention to the problem of extraterrestrial contamination. Specifically, scientists and aerospace engineers must control the terrestrial microbial contamination associated with any robot spacecraft intended to be in the vicinity, to fly by, to orbit, or to land on these interesting worlds. For future robot missions to Mars or Europa that will return materials from either of these worlds back to Earth for additional study, scientists and engineers must pay very close attention to the potential problem of back contamination of Earth. At present, NASA's policy for handling extraterrestrial samples returned to Earth is directed primarily toward containing any potentially hazardous material from Mars. Concerns expressed within the scientific community have included a difficult-to-control pathogen that is capable of infecting human hosts directly (generally considered to be highly unlikely) or the discovery of a life-form that is capable of upsetting the current natural balance of Earth's ecosystem.

It is also of paramount importance that responsible officials within the U.S. government, including NASA, carefully address—well in advance of any sample return mission—the potential public perception that might attribute some future epidemic, personal illness, or unusual event to a space-introduced contaminant from Mars. The public has already been exposed to and possibly preconditioned by a number of graphic science-fiction stories and motion pictures that center around the issue of a deadly disease (or creature) of extraterrestrial origin that spreads rapidly here on Earth, relentlessly killing large numbers of people. The classic "alien bug gone wild on Earth" story is the 1971 film, *The Andromeda Strain,* based on the excellent 1969 science-fiction novel of the same name by Michael Crichton. Other movies, such as *The Blob* and the set of *Alien* films, have vivid fictional presentations that reinforce the potential horrors of back contamination.

✧ Planetary Quarantine Program

NASA started a planetary quarantine program in the late 1950s at the beginning of the U.S. civilian space program. This quarantine program, conducted with international cooperation, was intended to prevent, or at least minimize, the possibility of contamination of alien worlds by early space probes. At that time, scientists were very concerned about the issue of forward contamination. In this type of extraterrestrial contamination, terrestrial microorganisms, "hitchhiking" on initial planetary probes and landers, could spread throughout another world, destroying any native life-forms, life precursors, or perhaps even remnants of past life-forms. If forward contamination occurred, it would compromise future scientific attempts to search for and identify extraterrestrial life-forms that had arisen independently of the Earth's biosphere. At the start of the 21st century, the concern about forward contamination remains high for missions to Mars and Europa.

To address potential extraterrestrial contamination problems, NASA scientists and engineers developed a rigorous planetary quarantine protocol. This protocol required that outbound unmanned planetary missions

Wearing proper protective clothing (to minimize the possibility of forward contamination), aerospace technicians inspect NASA's Mars exploration rover *Opportunity* in the Payload Hazardous Servicing Facility prior to launch from Cape Canaveral, Florida, on July 7, 2003. *Opportunity* successfully landed on Mars on January 25, 2004, and then began to explore the surface of the Red Planet in the Terra Meridiani region. *(NASA/JPL)*

be designed and configured to minimize the probability of alien-world contamination by terrestrial life-forms. As a design goal, these spacecraft and probes had a probability of 1 in 1,000 (1×10^{-3}) or less that they could contaminate the target celestial body with terrestrial microorganisms. Decontamination, physical isolation (for example, prelaunch quarantine), and spacecraft design techniques have all been used to support adherence to this protocol.

One simplified formula for describing the probability of planetary contamination is:

$$P(c) = m \times P(r) \times P(g)$$

where

P(c) is the probability of contamination of the target celestial body by terrestrial microorganisms,

m is the microorganism burden,

P(r) is the probability of release of the terrestrial microorganisms from the spacecraft hardware,

P(g) is the probability of microorganism growth after release on a particular planet or celestial object.

As previously stated, P(c) had a design goal value of less than or equal to 1 in 1,000. A value for the microorganism burden (m) was established by sampling an assembled spacecraft or probe. Then, through laboratory experiments, scientists determined how much this microorganism burden was reduced by subsequent sterilization and decontamination treatments. A value for P(r) was obtained by placing duplicate spacecraft components in simulated planetary environments. Unfortunately, establishing a numerical value for P(g) was a bit more tricky. The technical intuition of knowledgeable exobiologists and some educated "guessing" were blended together to create an estimate for how well terrestrial microorganisms might thrive on alien worlds that had not yet been visited. Of course, today, as scientists keep learning more about the environments on other worlds in humans' solar system, they can keep refining their estimates for P(g). Just how well terrestrial life-forms survive or possibly even grow on the Moon, Mars, Venus, Europa, Titan, and a variety of other interesting celestial bodies is the subject of future in-situ (on-site) laboratory experiments that will be performed by exobiologists or their robot spacecraft surrogates.

As a point of aerospace history, the early U.S. Mars flyby missions (for example, *Mariner 4,* launched on November 28, 1964, and *Mariner 6,* launched on February 24, 1969) had P(c) values ranging from 4.5×10^{-5} to 3.0×10^{-5}. These missions achieved successful flybys of the Red Planet on July 14, 1965, and July 31, 1969, respectively. Postflight calculations

Astronaut Charles Conrad, Jr., retrieves some equipment from the *Surveyor 3* robot spacecraft during the *Apollo 12* lunar landing mission in November 1969. Astronaut Alan L. Bean took this picture, and the lunar module *Intrepid* appears on the horizon in the right background. The *Surveyor 3* spacecraft made a soft landing on the Moon on April 19, 1967. The Apollo astronauts brought back some of the equipment from *Surveyor 3* so that NASA scientists, concerned with the issue of extraterrestrial contamination, could study the survival of hitchhiking terrestrial microorganisms after exposure to the harsh lunar environment for more than two and one-half years. The results were inconclusive. *(NASA)*

indicated that there was no probability of planetary contamination as a result of these successful precursor missions.

✧ Apollo Project Missions to the Moon

NASA's human-crewed Apollo Project missions to the Moon (1969–72) stimulated a great deal of debate about forward and back contamination. Early in the 1960s, scientists began to speculate in earnest: Is there life on the Moon? Some of the bitterest technical exchanges during the Apollo Project concerned this particular question. If there was life, no matter how primitive or microscopic, scientists wanted to examine it carefully and compare it with life-forms of terrestrial origin. This careful search for microscopic lunar life would, however, be very difficult and expensive because of the forward-contamination problem. For example, all equipment and materials landed on the Moon would need rigorous sterilization and decontamination procedures. There was also the glaring uncertainty about back contamination. If microscopic life did indeed exist on the Moon, scientists wondered whether such possible microscopic lunar life-forms would represent a serious hazard to the terrestrial biosphere. Because of the potential extraterrestrial-contamination problem, some members of the scientific community urged time-consuming and expensive quarantine procedures.

On the other side of this early 1960s contamination argument were those scientists (including the first generation of exobiologists) who emphasized the suspected extremely harsh lunar conditions: virtually no atmosphere; probably no water; extremes of temperature ranging from 248°F (120°C) at lunar noon to -238°F (-150°C) during the long, frigid lunar night; and unrelenting exposure to lethal doses of ultraviolet, charged particle, and X-ray radiations from the Sun, as well as very energetic cosmic rays from throughout the universe. No life-form, they argued, could possibly exist under such extremely hostile conditions.

This line of reasoning was countered by other scientists, who hypothesized that trapped water and moderate temperatures below the lunar surface might sustain very primitive life-forms. And so the great extraterrestrial-contamination debate raged back and forth until finally the *Apollo 11* expedition departed on the first lunar-landing mission. As a compromise, the *Apollo 11* mission flew to the Moon with careful precautions against back contamination but with only a very limited effort to protect the Moon from forward contamination by terrestrial organisms.

The Lunar Receiving Laboratory (LRL) at the Johnson Space Center in Houston provided quarantine facilities for two years after the first lunar landing. What scientists learned during its operation serves as a useful starting point for planning new Earth-based or space-based quarantine facilities. In the future, advanced quarantine facilities will be needed to accept, handle, and test extraterrestrial materials from Mars and other solar-system bodies of interest in our search for alien life-forms (present or past).

During the Apollo Project, no evidence was discovered that native alien life was then present or had ever existed on the Moon. Scientists at the Lunar Receiving Laboratory performed a careful search for carbon, since terrestrial life is carbon based. Found in the lunar samples were 100 to 200 parts per million of carbon. Of this amount, only a few tens of parts per million are considered indigenous to the lunar material, while the bulk amount of carbon has been deposited by the solar wind. Exobiologists and lunar scientists have concluded that none of this carbon appears derived from biological activity. In fact, after the first few Apollo expeditions to the lunar surface, even back-contamination quarantine procedures of isolating the Apollo astronauts for a period of time were dropped after the *Apollo 14* mission.

Has the life-on-the-Moon debate ended? Quite possibly it has not because of discoveries made within the last decade or so. The suspected presence of lunar water ice in permanently shadowed craters found in the polar regions of the Moon may revive some very modest portion of the "microscopic lunar life" debate of the early 1960s. Furthermore, here on Earth, scientists have discovered various extremophiles—very hardy microorganisms that are capable of living in extremely harsh environmental conditions. If these hardy life-forms have been discovered in the strangest places on Earth, what about elsewhere in the solar system in places where there is water?

✧ *Apollo 11*—The First Visit by Humans to Another World

The *Apollo 11* mission achieved the national goal set by President Kennedy in 1961—namely landing human beings on the surface of the Moon and returning them safely to Earth within the decade of the 1960s. On July 20, 1969, astronauts Neil Armstrong (commander) and Edwin (Buzz) Aldrin (lunar module pilot) flew the lunar module (called *Eagle*) to the surface of the Moon, touching down safely in the Sea of Tranquility. While Armstrong and then Aldrin became the first two persons to walk on another world, their fellow astronaut Michael Collins (the CM pilot) orbited above in the *Columbia* command and service module (CSM).

On July 16, 1969, a gigantic Saturn V rocket flawlessly lifted off its pad at Complex 39-A of the Kennedy Space Center and started the most profound journey of exploration in human history. The *Apollo 11* spacecraft (CSM-LM combined) followed a similar mission profile to the Moon, as had its immediate predecessor, the *Apollo 10* spacecraft. Following launch, *Apollo 11* entered orbit around Earth. After completing one and one-half orbits of Earth, the third stage of the Saturn V (the S-IVB upper stage) reignited for an approximately six-minute-duration translunar injection burn that placed the spacecraft on a course for the Moon. Thirty-three minutes later, the *Apollo 11* astronauts separated the *Columbia* CSM from the S-IVB stage, turned the spacecraft around, and docked with the *Eagle* LM. About 75 minutes later, they released the S-IVB and injected this spent rocket stage into heliocentric orbit. While the docked CSM-LM spacecraft configuration coasted through cislunar space for its rendezvous with history, the *Apollo 11* astronauts made a live, color, television broadcast back to Earth.

On July 19, the astronauts performed a 358-second, retrograde firing of the CSM's service propulsion system (SPS) to achieve insertion into lunar orbit. This orbit insertion burn was accomplished while the spacecraft was behind the Moon and out of contact with Earth. A second, much shorter duration (17-second) SPS burn circularized the spacecraft's orbit around the Moon.

The next day (July 20), Armstrong and Aldrin entered the *Eagle* to perform a final checkout before traveling in the LM down to the lunar surface. As the LM and CSM separated, Collins made a visual inspection of the *Eagle* from the *Columbia* (CSM). Armstrong and Aldrin then fired the LM's descent engine for 30 seconds. Their actions put the *Eagle* in a descent orbit, which had a closest approach to the Moon's surface of nine miles (14.5 km). The two astronauts fired the LM descent engine once again, this time for 756 seconds, and they began the final phase of their historic descent to the Moon's surface. Although Armstrong piloted the LM to a safe touchdown on the lunar surface, when he finished he had less than 30 seconds of propellant supply remaining. The problem that the astronauts encountered was finding a suitable landing site. Despite all the previous photographic reconnaissance that was performed during the site-selection process, the original landing site chosen in Mare Tranquilitatis (the Sea of Tranquility) was actually populated with a large number of small craters and rocks—landing incorrectly near any one of which could have spelled disaster for the mission. Searching for a suitable landing spot as the fuel supply for the *Eagle*'s descent engine approached exhaustion, Armstrong finally spotted a relatively smooth place and quickly set the spidery-looking spacecraft down at 4:17 P.M. (EDT) on July 20, 1969.

Throughout this harrowing search for a safe lunar landing spot, personnel at NASA's mission controller center in Houston, Texas, were very anxiously monitoring the depletion of the *Eagle*'s propellant supply. As soon as signals from the LM indicated that some type of contact had been made with the surface, mission control sent the following short message: "We copy you down, *Eagle*." The modest time delay (a little more than two seconds) for radio signals to go back and forth between Earth and the Moon seemed like an eternity to everyone in the room that day. Then back came Armstrong's famous reply: "Houston, Tranquility Base here. The *Eagle* has landed!" The response from Houston at this historic moment proved equally memorable: "Roger, Tranquility. We copy you on the ground. You've got a bunch of guys here about to turn blue. We're breathing again. Thanks a lot." This simple dialogue marked the start of one of the greatest moments in exploration.

Six hours later, Armstrong opened the ingress-egress hatch on the LM and cautiously descended the ladder. As his left foot made contact with the lunar soil, he reported back to Houston: "That's one small step for (a) man . . . one giant leap for mankind." About 19 minutes later, Aldrin followed and became the second human being to walk on the Moon. As he looked out at the lunar landscape and noticed the starkness of the shadows and the barren, almost desertlike characteristics of the Moon's surface, Aldrin remarked: "Beautiful, beautiful. Magnificent desolation."

The important deed was done. Intelligent life had successfully traveled from the surface of one world in this solar system to the surface of another. Consciousness was physically starting to spread beyond the boundaries of Earth. Now the human race faced a very interesting societal decision: Will it be the galaxy or nothing (self-annihilation)?

Like tourists everywhere, Armstrong and Aldrin began their visit to the Moon by snapping pictures, lots of pictures. They also collected souvenirs, some 47.7 pounds (21.7 kg) of soil and rock samples for the planetary scientists back home on Earth. Once their initial euphoria subsided a bit, they began to deploy instruments, such as the Early Apollo Surface Experiments Package (EASEP) near the lunar module. In a very busy few hours, they traversed a total of about 820 feet (250 meters) across the Moon's surface, gathered rock specimens, inspected the LM, positioned science instruments, and planted an American flag—not as a symbol of territorial claim (an act prohibited by international treaty) but rather as a permanent symbol of the nation that accomplished the first human landing. They also removed the protective, thin metal plate that was covering the *Apollo 11* plaque mounted on the LM's ladder.

At the conclusion of their extravehicular activity (EVA), the astronauts returned to the LM and closed the hatch. They were supposed to sleep for a few hours before attempting to blast off from the lunar surface and rejoin Michael Collins, who was orbiting above in the *Columbia* CSM. Clearly, there was far too much to do and see and so little time. The best "rest" Aldrin managed to accomplish on the floor of the cramped lunar module was (in his own words) a "couple of hours of mentally fitful drowsing." Armstrong simply stayed awake inside the tiny crew cabin that was filled with noisy pumps and bright warning lights.

On July 21, after spending 21 hours and 36 minutes on the lunar surface, the astronauts fired the LM's ascent engine and lifted off from the Moon's surface. As the upper half of the *Eagle* arose into lunar orbit, the lower half remained behind on the surface in the Sea of Tranquility at 0.67 degrees N latitude, 23.5 degrees E longitude (lunar coordinates). The *Eagle* (as well as the other abandoned lunar-module descent stages from *Apollo 12, 14, 15, 16,* and *17*) now serves as a permanent memorial to humans' conquest of space. Armstrong and Aldrin then docked with *Columbia* and transferred the collection of lunar rocks and some equipment into the command module.

On July 22, in preparation for the journey back to Earth, the astronauts jettisoned the ascent stage of the lunar module into orbit around the Moon. The precise fate of the upper half of the *Eagle* LM is not known, but NASA mission managers assumed that it crashed into the Moon's surface within a month to four months after being abandoned in orbit. After completing 31 revolutions of the Moon, the *Columbia* prepared to return home to Earth. A two-and-one-half-minute firing of the CSM's main rocket engine began the all-important transearth injection process.

On the morning of July 24, the command module made its programmed separation from service module (SM) and its three occupants prepared for reentry. After a mission elapsed time of 195 hours, 18 minutes, and 35 seconds, Armstrong, Aldrin, and Collins splashed down in the middle of the Pacific Ocean about 15 miles (24 km) away from the recovery ship, USS *Hornet*. U.S. Navy recovery crews arrived quickly by helicopter and tossed biological isolation garments into the spacecraft. After the suitably "cocooned" astronauts emerged from the *Columbia* command module, the team of recovery swimmers swabbed the spacecraft's hatch down with an organic iodine solution. Then the astronauts and recovery team personnel decontaminated each other's protective garments with a solution of sodium hypochlorite. The three astronauts were then plucked from the ocean's surface, transported by helicopter to the USS *Hornet*, and placed immediately in a special lunar quarantine trailer facility on the deck of the aircraft carrier. After a quick change of clothing inside the quarantine facility, the astronauts appeared at the window and received personal congratulations from President Richard M. Nixon (who had flown to the USS *Hornet*).

While confined in their quarantine trailer, Armstrong, Aldrin, and Collins—along with their now biologically isolated command module and its precious cargo of lunar rocks—traveled (primarily by aircraft) to the Lunar Receiving Laboratory in Houston, Texas. There, they remained in quarantine until late in the evening on August 10. From a medical perspective, this period of quarantine proved totally uneventful for the astronauts. Showing no signs of any ill effects from exposure to lunar dust or to any "postulated" extraterrestrial microorganism that might have hitchhiked back to Earth on their spacesuits or equipment, NASA's biomedical experts decided to release the three astronauts from quarantine. They went home to their families and, after a much-deserved period of privacy, embarked on a triumphant tour around the world.

To demonstrate that NASA scientists harbor no lingering concerns about back contamination from the Moon, there is a publicly accessible lunar rock sample and the *Apollo 11* command module (*Columbia*) on

Though separated by the window of the Mobile Quarantine facility (located on the deck of the USS *Hornet*), President Richard M. Nixon was still able to share a humorous moment with *Apollo 11* astronauts (left to right) Neil A. Armstrong, Michael Collins, and Edwin (Buzz) Aldrin shortly after they splashed down in the Pacific Ocean on July 24, 1969. Uncertain about the issue of back contamination, NASA managers quarantined the crews of the first three successful lunar landing missions (*Apollo 11, 12,* and *14*). When no harmful effects appeared, NASA abandoned these quarantine procedures for the crews of the remaining lunar landing missions (*Apollo 15, 16,* and *17*). *(NASA)*

exhibit at the National Air and Space Museum in Washington, D.C. Almost all the lunar rocks brought back by the Apollo astronauts are now kept under strictly controlled environmental conditions to preserve their scientific research value.

✧ Preventing Back Contamination

There are three fundamental approaches toward handling extraterrestrial samples to avoid back contamination. First, scientists could sterilize a sample while it is en route to Earth from its native world. Second, they could place it in quarantine in a remotely located, maximum-confinement facility on Earth while scientists examine it closely. Finally, they could also perform a preliminary hazard analysis (called the extraterrestrial protocol tests) on the alien sample in an orbiting quarantine facility before they allow the sample to enter the terrestrial biosphere. To be adequate, a quarantine facility must be capable of (1) containing all alien organisms present in a sample of extraterrestrial material, (2) detecting these alien organisms during protocol testing, and (3) controlling these organisms after detection until scientists could dispose of them in a safe manner.

One way to bring back an extraterrestrial sample that is free of potentially harmful alien microorganisms is to sterilize the material during its flight to Earth. However, the sterilization treatment used must be intense enough to guarantee that no microbial life-forms as exobiologists currently know or anticipate them could survive. An important concern here is also the impact that the sterilization treatment might have on the scientific value of the alien world sample. For example, use of chemical sterilants would most likely result in contamination of the sample, thereby preventing the measurement of certain soil properties. Heat could trigger violent chemical reactions within the soil sample, resulting in significant changes and the loss of important exogeological data. Finally, sterilization would also greatly reduce the biochemical information content (if any) within the sample. It is quite questionable as to whether any useful exobiology data can be obtained by analyzing a heat-sterilized alien material sample. To put it simply, in their search for extraterrestrial life-forms, exobiologists want "virgin alien samples."

If scientists decide not to sterilize the alien samples en route to Earth, they then have only two general ways to avoid potential back contamination problems. First, they can deliver to and secure the unsterilized sample of alien material in a maximum quarantine facility on Earth. The scientists would then conduct detailed scientific investigations on the sample in this isolated terrestrial laboratory. The second choice is for the scientists to intercept and inspect the sample at an orbiting quarantine facility (OQF) before allowing the material to enter Earth's biosphere.

The technology and procedures for hazardous-material containment have been used on Earth in the development of highly toxic chemical- and biological-warfare agents and in conducting research involving highly infectious diseases. A critical question for any quarantine system is whether the containment measures are adequate to hold known or suspected pathogens while experimentation is in progress. Since the characteristics of potential alien organisms are not presently known, scientists must assume that the hazard such (hypothetical) microorganisms could represent is at least equal to that of terrestrial Class IV pathogens. (A terrestrial Class IV pathogen is an organism that is capable of being spread very rapidly among humans: No vaccine exists to check its spread; no cure has been developed for it; and the organism produces high mortality rates in infected persons.)

Judging from the large uncertainties associated with potential extraterrestrial life-forms, it is not obvious that any terrestrial quarantine facility

GENESIS SOLAR WIND SAMPLE RETURN MISSION

The primary mission of NASA's *Genesis* spacecraft was to collect samples of solar wind particles and return these samples of extraterrestrial material safely to Earth for detailed analysis. The mission's specific science objectives were to obtain precise solar isotopic and elemental abundances and to provide a reservoir of solar matter for future investigation. A detailed study of captured solar wind materials would allow scientists to test various theories of solar-system formation. Access to these materials would also help them resolve lingering issues about the evolution of the solar system and the composition of the ancient solar nebula.

The mission started on August 8, 2001, when an expendable Delta II rocket successfully launched the 1,400-pound (636-kg) *Genesis* spacecraft from Cape Canaveral Air Force Station, Florida. Following launch, the cruise phase of the mission lasted slightly more than three months. During this period, the spacecraft traveled 932,000 miles (1.5 million km) from Earth to the Lagrange libration point 1 (L1). The *Genesis* spacecraft entered a halo orbit around L1 on November 16, 2001. On arrival, the spacecraft's large thrusters fired, putting *Genesis* into a looping, elliptical orbit around the Lagrangian point. The *Genesis* spacecraft then completed five orbits around L1; nearly 80 percent of the mission's total time was spent collecting particles from the Sun.

On December 3, 2001, *Genesis* opened its collector arrays and began to accept particles of solar wind. A total of 850 days were logged, exposing the special collector arrays to the solar wind. These collector arrays are circular trays composed of palm-sized hexagonal tiles made of various high-quality materials, such as silicon, gold, sapphire, and diamondlike carbon. After the sample return capsule opened, the lid of the science canister opened as well, exposing a collector for the bulk solar wind. As long as the science canister's lid was opened, this bulk collector array was exposed to different types of solar wind that flowed past the spacecraft.

The other dedicated science instrument of the *Genesis* spacecraft was the solar wind collector. As its name implies, this instrument would concentrate the solar wind onto a set of small collector tiles made of diamond, silicon carbide, and diamondlike carbon. As long as the lid of the science canister was opened, the concentrator was exposed to the solar wind throughout the collection period.

On April 1, 2004, ground controllers ordered the robot spacecraft to stow the collectors and so its collection of pristine particles from the Sun ended. The closeout process was completed on April 2, when the *Genesis* spacecraft closed and sealed its sample return capsule. Then, on April 22, the spacecraft began its journey back toward Earth. However, because of the position of the landing site—the U.S. Air Force's Utah Testing and Training Range in the northwestern corner of that state—and the will gain very wide acceptance by the scientific community or the general public. For example, locating such a facility and all its workers in an isolated area on Earth actually provides only a small additional measure of protection. Consider the planetary environmental-impact controversies that could rage as individuals speculated about possible ecocatastrophes.

unique geometry of the *Genesis* spacecraft's flight path, the robot sampling craft could not make a direct approach and still make a daytime landing. To allow the *Genesis* chase-helicopter crews an opportunity to capture the return capsule midair in daylight, the *Genesis* mission controllers designed an orbital detour toward another Lagrange point, L2, located on the other side of Earth from the Sun. After successfully completing one loop around L2, the *Genesis* spacecraft was prepared for its return to Earth on September 8. On September 8, the spacecraft approached Earth and performed a number of key maneuvers prior to releasing the sample return capsule. Sample capsule release took place when the spacecraft flew past Earth at an altitude of about 41,000 miles (66,000 km). As planned, the *Genesis* return capsule successfully reentered Earth's atmosphere at a velocity of 6.8 miles per second (11 km/s) over northern Oregon.

Unfortunately, during reentry on September 8, the parachute on the *Genesis* sample return capsule failed to deploy (apparently because of an improperly installed gravity switch), and the returning capsule smashed into the Utah desert at a speed of 193 miles per hour (311 km/hr). The high-speed impact crushed the sample return capsule and breeched the sample collection capsule—possibly exposing some collected pristine solar materials to potential contamination by the terrestrial environment.

However, mission scientists worked diligently to recover as many samples as possible from the spacecraft wreckage and then to ship the recovered materials in early October to the Johnson Space Center in Houston, Texas, for evaluation and analysis. One of the cornerstones of the recovery process was the discovery that the gold foil collector was undamaged by the hard landing. Another postimpact milestone was the recovery of the *Genesis* spacecraft's four separate segments of the concentrator target. Designed to measure the isotopic ratios of oxygen and nitrogen, the segments contain within their structure the samples that are the mission's most important science goals.

The *Genesis* sample of extraterrestrial materials consisted of atoms collected from the Sun. NASA's planetary protection officer had categorized the mission as "safe for unrestricted Earth return." This declaration meant that exobiologists and other safety experts had concluded that there was no chance of extraterrestrial contamination during sample collection at the L1. The U.S. National Research Council's Space Studies Board also concurred on the planetary protection designation of unrestricted Earth return for the Genesis mission. The board determined that the sample had no potential for containing life. Consequently, there was no significant issue of extraterrestrial contamination of planet Earth due to the aborted sample return operation of the *Genesis* capsule. However, the issue of planetary contamination remains of concern when robot-collected sample capsules return from potentially life-bearing celestial bodies, like Mars and Europa.

What would happen to life on Earth if alien organisms did escape and went on a deadly rampage throughout the Earth's biosphere? As mentioned previously, dangerous alien microorganisms that have gone wild in the terrestrial biosphere has been a popular and recurrent science-fiction theme. Although causing no back contamination hazard, the unplanned, high-speed crash of NASA's *Genesis* spacecraft sample return capsule into the Utah desert on September 8, 2004, helped reinforce legitimate concern that if something might go wrong during a complex sample return mission—it can and often will. The alternative to this potentially explosive controversy is quite obvious: Locate the quarantine facility in outer space.

✦ Orbiting Quarantine Facility

A space-based, orbiting quarantine facility (OQF) provides several distinct advantages. First, the OQF eliminates the possibility of a sample return spacecraft crashing and accidentally releasing a potentially deadly cargo of alien microorganisms. Second, the space-based facility guarantees that any alien organisms that might escape from confinement systems within the orbiting complex cannot immediately enter Earth's biosphere. Finally, the postulated OQF would ensure that all quarantine workers remain in total physical isolation during protocol testing.

As robot spacecraft and crewed vehicles expand the human sphere of influence into heliocentric space, people must also remain vigilant about the potential hazards of extraterrestrial contamination. Scientists, space explorers, and extraterrestrial entrepreneurs must be aware of the ecocatastrophes that might occur when "alien worlds collide"—especially on the microorganism level.

With a properly designed and operated orbiting quarantine facility, alien-world materials can be tested for potential hazards. Three hypothetical results of such protocol testing are: (1) No replicating alien organisms are discovered; (2) replicating alien organisms are discovered, but they are also found not to be a threat to terrestrial life-forms; or (3) hazardous replicating alien life-forms are discovered. If potentially harmful (hypothetical) replicating alien microorganisms are discovered during these protocol tests, then quarantine facility workers would either render the sample harmless (for example, through heat- and chemical-sterilization procedures); retain it under very carefully controlled conditions in the orbiting complex and perform more detailed analyses on the alien life-forms; or properly dispose of the sample before the alien life-forms could enter Earth's biosphere and infect terrestrial life-forms. If alien microorganisms infected any of the workers on the orbiting quarantine facility, the scientists would be kept isolated at the OQF and treated there. Under no

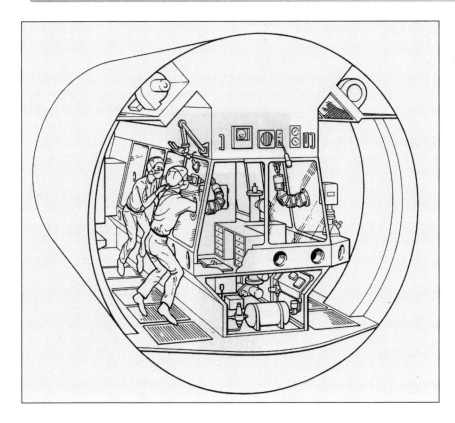

Exobiologists in an orbiting quarantine facility (OQF) test for potentially harmful alien microorganisms that might be contained in an extraterrestrial soil sample. (NASA)

circumstances would an infected human worker be allowed to return to Earth from the OQF.

✧ Contamination Issues in Exploring Mars

Increasing interest in Mars exploration has also prompted a new look at the planetary protection requirements for forward contamination. In 1992, for example, the Space Studies Board of the U.S. National Academy of Sciences recommended changes in the requirements for Mars landers that significantly alleviated some of the expense and operational burden of planetary protection for these missions. The board's recommendations were published in the document "Biological Contamination of Mars: Issues and Recommendations" and were presented at the 29th COSPAR Assembly that was held in 1992 in Washington, D.C. In 1994, a resolution addressing these recommendations was adopted by COSPAR at the 30th assembly and subsequently has been incorporated into NASA's planetary protection policy. Of course, as scientists learn more about Mars, planetary protection requirements will be adjusted to reflect current scientific knowledge and to achieve the overall objective of

BIOCLEAN ROOM

A bioclean room, or clean room, is any enclosed area where there is control over viable and non-viable particulates in the air with temperature, humidity, and pressure control as required to maintain specified standards. A viable particle is a particle that will reproduce to form observable colonies when placed on a specified culture medium and when incubated according to optimum environmental conditions for growth after collection. A nonviable particle will not reproduce to form observable colonies when placed on a specified culture medium and when incubated according to optimum environmental conditions for growth after collection.

Scientists and engineers operate a bioclean room with emphasis on minimizing airborne viable and nonviable particle generation or concentrations to specified levels. Particle size is expressed as the apparent maximum linear dimension or diameter of the particle—usually in micrometers (μm) or microns. (A micron is one-millionth of a meter.)

There are three clean-room classes that are based on total (viable and nonviable) particle count with the maximum number of particles per unit volume or horizontal surface area being specified for each class. The cleanest, or most stringent, airborne particulate environment is called a Class 100 Bioclean Room (or a Class 3.5 Room in SI units). In this type of bioclean room, the particle count may not exceed a total of 100 particles per cubic foot (3.5 particles per liter) of

protecting any native life-forms on Mars, no matter how humble, from contamination by terrestrial microorganisms.

These new recommendations recognize the very low probability of growth of (terrestrial) microorganisms on the Martian surface. With this assumption in mind, the forward contamination protection policy shifts from probability of growth considerations to a more direct and determinable assessment of the number of microorganisms with any landing event. For landers that do not have life-detection instrumentation, the level of biological cleanliness required is that of the Viking Project spacecraft prior to heat sterilization. Class 100,000 clean-room assembly and component testing can accomplish this level of biological cleanliness. This is considered a very conservative approach that minimizes the chance of compromising future exploration. Landers with life-detection instruments would be required to meet Viking Project spacecraft poststerilization levels of biological cleanliness or levels driven by the search-for-life experiment itself. Scientists recognize that the sensitivity of a life-detection instrument may impose the more severe biological cleanliness constraint on a Mars lander mission.

Included in recent changes to COSPAR's planetary protection policy is the option that an orbiter spacecraft not be required to remain in orbit around Mars for an extended time if it can meet the biological cleanliness standards of a lander without life-detection experiments. In addition, the a size 0.5 micrometer (μm) and larger; the viable particle count may not exceed 0.1 per cubic foot (0.0035 per liter) with an average value not to exceed 1,200 per square foot (12,900 per square meter) per week on horizontal surfaces. A Class 10,000 Bioclean Room (or a Class 350 Room in SI units) is the next-cleanest environment, followed by a Class 100,000 Bioclean Room (or a Class 3,500 Room in SI units).

In aerospace applications, bioclean rooms are used to manufacture, assemble, test, disassemble, and repair delicate spacecraft sensor systems, electronic components, and certain mechanical subsystems. Aerospace workers wear special protective clothing, including gloves, smocks (frequently called bunny suits), and head and foot coverings to reduce the level of dust and contamination in clean rooms. A Class 10,000 Bio-clean Room facility is typically the cleanliness level encountered in the assembly and testing of large spacecraft. If a planetary probe is being prepared for a trip to a possible life-bearing planet such as Mars, care is also taken in the bioclean room to ensure that "hitchhiking terrestrial microorganisms" are brought to the minimum population levels consistent with planetary quarantine protocols—thereby avoiding the potential of forward contamination of the target alien world.

Planetary scientists and exobiologists will place extraterrestrial soil and rock samples that are possibly life-bearing in highly isolated and highly secure bioclean rooms. This will help to avoid any possible back contamination of the terrestrial biosphere as a result of space exploration missions involving sample returns from alien worlds.

probability of inadvertent early entry (into the Martian atmosphere) has been relaxed compared to previous requirements.

The present policy for samples returned to Earth remains directed toward containing potentially hazardous Martian material. Concerns still include a difficult-to-control pathogen that is capable of directly infecting human hosts (currently considered extremely unlikely) or a life-form that is capable of upsetting the current ecosystem. Therefore, for a future Mars Sample Return Mission (as discussed in chapter 4), the following backward contamination policy now applies. All samples would be enclosed in a hermetically sealed container. The contact chain between the return space vehicle and the surface of Mars must be broken to prevent the transfer of potentially contaminated surface material by means of the return spacecraft's exterior. The sample would be subjected to a comprehensive quarantine protocol to investigate whether or not harmful constituents are present. It should also be recognized that even if the sample return mission has no specific exobiological goals, the mission would still be required to meet the planetary protection sample return procedures as well as the life-detection protocols for forward contamination protection. This policy not only mitigates concern of potential contamination (forward or back), but it also prevents a hardy terrestrial microorganism "hitchhiker" from masquerading as a Martian life-form.

✧ Protecting Europa from Forward Contamination

Extensive observations of Europa by NASA's *Galileo* spacecraft (see chapter 5) indicate that this large moon of Jupiter has been geologically active in the relatively recent past and that liquid water could exist beneath its surface shell of water ice—a moon-encircling shell that appears to be between 6 miles (10 km) and 106 miles (170 km) thick.

Even though scientists do not have sufficient information at this time to state with confidence that Europa has a subsurface, liquid-water ocean, native (extraterrestrial) life, or supports environments that might prove compatible with terrestrial life (at the microbial level), such intriguing possibilities cannot be quickly dismissed. Therefore, consistent with international planetary-body-protection agreements, all future robot spacecraft missions to Europa must be subjected to rigorous procedures to prevent forward contamination by hitchhiking terrestrial microorganisms. Unless steps are taken to prevent Europa's contamination by terrestrial microorganisms, the future scientific integrity of the search for alien life on this Jovian moon will be severely compromised.

The discovery of extremophiles—terrestrial organisms living in extreme environments—has caused scientists to recognize the uncertainties in their knowledge of the diversity of life on their own home planet. So care must be taken to avoid introducing terrestrial microorganisms to the surface or subsurface of Europa. If this moon has an extensive, perhaps global, subsurface ocean, then the accidental introduction of a viable microorganism from Earth could result in a rapid spread of forward contamination throughout the entire aquatic domain. Any native Europan life-forms could be overwhelmed quickly by the microbial invaders from Earth and soon could become extinct, long before scientists had a chance (through surrogate robot spacecraft missions) to study them. Consequently, national and international planetary protection specialists are currently recommending that for every future mission to Europa, the probability of the robot spacecraft contaminating a Europan ocean with a viable terrestrial organism at any time in the future (that is, to the next 100 years or so) should be less than 10^{-4} per mission.

As part of its suggested planetary protection guidelines for Europa, COSPAR allows aerospace engineers and spacecraft designers to take advantage of the intense ionizing radiation environment encountered in the Jovian environment to help reduce the bioload on any robot spacecraft visiting Europa. One of NASA/JPL's radiation dose models for Europa indicates that at a 3.9-inch (10-cm) depth in Europan ice, the natural ionizing radiation dose is approximately 5,000 rad (50 gray) per month. By way of comparison, the natural ionizing radiation environment at the surface of Earth is on the order of 0.1 rad (0.001 gray) per year. Similarly, the ionizing dose limits (wholebody) for astronauts is 25 rads (0.25 gray) per month and 50 rad (0.5 gray) per year, not to exceed a career total of between 100 rad (1 gray) to 400 rad (4 gray), depending on age and sex. Even a brief visit to the surface of Europa would prove fatal to an astronaut, due to acute radiation syndrome. So in all likelihood, only radiation-hardened, well-sterilized robot exploring machines will search for life in any (suspected) Europan ocean later this century.

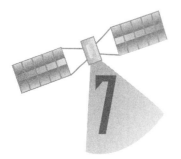

Habitable Worlds around Distant Stars

An *extrasolar planet* is a planet that belongs to a star other than the Sun. Modern astronomers use two general methods to detect extrasolar planets: direct—involving a search for telltale signs of a planet's infrared emissions—and indirect—involving precise observation of any perturbed motion (for example, wobbling) of the parent star or any periodic variation in the intensity or spectral properties of the parent star's light.

Growing evidence of planets around other stars is helping astronomers to validate the hypothesis that planet formation is a normal part of stellar evolution. Detailed physical evidence concerning extrasolar planets—especially if scientists can determine their frequency of occurrence as a function of the type of star—would greatly assist scientists in estimating the cosmic prevalence of life. If life originates on suitable (Earth-like) planets whenever it can (as many exobiologists currently hold), then knowing how abundant such suitable planets are in the Milky Way would allow scientists to make more credible decisions (that is, better educated guesses) about where to search for extraterrestrial intelligence and the probability of finding life (intelligent or otherwise) beyond humans' solar system.

Starting in the 1990s, scientists began to detect (through spectral light variations of the parent star) Jupiter-sized planets around Sunlike stars, such as 51 Pegasi, 70 Virginis, and 47 Ursae Majoris. Detailed computer analyses of spectrographic data have revealed that light from these stars appears redder and then bluer in an alternating periodic (sine wave) pattern. This periodic light pattern could indicate that the stars themselves are moving back and forth along the line of sight possibly due to a large (unseen) planetary object that is slightly pulling the stars away from (redder spectral data) or toward (bluer spectral data) Earth. The suspected planet around 51 Pegasi is sometimes referred to as a hot Jupiter, since it appears to be a large planet (about half of Jupiter's mass) that is located so

This is an artist's rendering of a large extrasolar planet, known as a hot Jupiter, orbiting around an alien star. This particular gas–giant planet orbits the yellow Sunlike star HD 209458, which is 150 light–years away from Earth. Astronomers used NASA's *Hubble Space Telescope (HST)* to look at this world and to make the first detection of an atmosphere around an extrasolar planet. The planet was not directly seen by the *HST*. Instead, the presence of sodium was detected in light that was filtered through the planet's atmosphere when it passed in front of its parent star, as seen from Earth—an event called a transit. The planet was discovered in 1999 by its subtle gravitational pull on the star. The planet has about 70 percent the mass of Jupiter and orbits the star at a distance of only 4 million miles (6.4 million km). *(NASA and Greg Bacon [STScI/AVL])*

close to its parent star that it completes an orbit in just a few days (approximately 4.23 days). The suspected planetary body orbiting 70 Virginis lies about one-half an astronomical unit (AU) distance from the star and has a mass approximately eight times that of Jupiter. Finally, the object orbiting 47 Ursae Majoris has an estimated mass that is 3.5 times that of Jupiter. It orbits the parent star at approximately AUs distance, taking about three years to complete one revolution.

To find extrasolar planets and characterize their atmospheres, scientists will use current (or planned) space-based observatories like the *Spitzer Space Telescope (SST)*, the *James Webb Space Telescope (JWST)*, and the *Kepler* spacecraft. In late 2003, NASA's *Spitzer Space Telescope* captured a dazzling image of a massive disc of dusty debris encircling a nearby star called Fomalhaut. Planetary scientists consider such discs as remnants of planetary construction and believe that Earth formed out of a similar disc.

The *Spitzer Space Telescope* is now helping scientists identify other stellar dust clouds that might mark the sites of developing planets. In 2004, this space-based infrared telescope gathered data to indicate a possible planet spinning its way through a clearing in a nearby star's dusty, planet-forming disc. *Spitzer* detected this clearing around the star CoKu Tau 4. Astronomers believe than an orbiting massive body, like a planet, may have swept away the star's disc material, leaving a central hole. The possible planet is theorized to be at least as massive as Jupiter and may have a similar appearance to what the giant planets in this (humans') solar system looked like billions of years ago. As shown in the artist's rendering, a graceful set of rings, much like Saturn's, spin high above the planet's cloudy atmosphere. The set of rings is formed from countless small orbiting particles of dust and ice—leftovers from the initial gravitational collapse that formed the suspected giant planet.

This is an artist's rendering of a suspected extrasolar planet. In May 2004, a clearing was detected around the star CoKu Tau 4 by NASA's *Spitzer Space Telescope.* Astronomers believe that an orbiting massive body (like the planet depicted here) may have swept away the star's disc material, leaving a central hole. *(NASA/JPL)*

If some day human beings were able to visit an extrasolar planet like this one, they would have a very different view of the universe. The sky, instead of being the familiar dark expanse lit by distant stars, would be dominated by the thick disc of dust that fills the young planetary system. The view looking toward the alien solar system's parent star (CoKu Tau 4) would be relatively clear because the dust in the interior disk has already fallen into the accreting star. A bright band would appear to surround the central star, caused by the central star's light being scattered back by the dust in the disc. Looking away from CoKu Tau 4, the dusty disc would appear dark, blotting out light from all the stars in the sky except those that lie well above the plane of the disc.

The *James Webb Space Telescope (JWST)* will be a large single telescope that is folded to fit inside its launch vehicle and cooled to low temperatures in deep space to enhance its sensitivity to faint, distant objects. Mission controllers will operate *JWST* in an orbit far from Earth, away from the thermal energy (heat) radiated to space by humans' home planet. It is scheduled for launch in 2011, and one of this observatory's main science goals is to determine how planetary systems form and interact. The *JWST* will be able to observe evidence of the formation of planetary systems (some of which may be similar to this solar system) by mapping the light from the clouds of dust grains orbiting stars.

The star, Beta Pictoris, has such a cloud, as was discovered by the Infrared Astronomical Satellite (IRAS). These clouds are bright near the host stars and may be divided into rings by the gravitational influence of large planets. Scientists speculate that this dust represents material forming into planets. Around older stars, such dust clouds may be the debris of material that failed to condense into planets. The *JWST* will have unprecedented sensitivity to observe faint dust clouds around nearby stars. The infrared wavelength range is also the best way to search for planets directly because they are brighter, relative to their central stars. For example, at visible wavelengths, Jupiter is about 100 million times fainter than the Sun, but in the infrared, it is only 10,000 times fainter. A planet like Jupiter would be difficult to observe directly with any telescope on Earth, but the *JWST* has a chance—because it operates in space beyond the disturbing influences of the terrestrial atmosphere.

Currently scheduled for launch in November 2008, NASA's *Kepler* spacecraft will use a unique space-based telescope that has been specifically designed to search for Earth-like planets around stars beyond the solar system. The *Kepler* spacecraft will allow scientists to search the Milky Way galaxy for Earth-size or even smaller planets. The extrasolar planets discovered thus far are giant planets, similar to Jupiter, and are probably composed mostly of hydrogen and helium. So exobiologists believe that such Jupiter-sized planets are unlikely to harbor life. However, some

This artist's rendering shows the view from a hypothetical moon in orbit around the first known planet to reside in a tight-knit triple star system. The planet, called HD 188753 Ab, is a gas giant with about 1.14 times the mass of Jupiter and an orbital period of 3.3 days. The giant planet travels around a single star that is orbited by a pair of pirouetting stars. The planet with three suns was discovered using the Keck I telescope atop Mauna Kea in Hawaii. The triple star family is called HD 188753 and is located 149 light-years away in the constellation Cygnus. *(NASA/JPL–Caltech)*

scientists have suggested that these massive planets may have moons with an atmosphere and liquid water on the surface. In such cases, life could arise in these planetary systems. The Kepler mission is especially important because none of the extrasolar planet detection methods used to date have had the capability of finding Earth-sized planets—that is, planets that are 30 to 600 times less massive than Jupiter. Furthermore, none of the giant extrasolar planets discovered to date probably has not had liquid water on its surface or even a solid surface.

The *Kepler* spacecraft is different from previous ways of looking for planets because it will look for the transit signature of planets. A transit occurs each time a planet crosses the line-of-sight between the planet's parent star that it is orbiting and the observer. When this happens, the planet blocks some of the light from its star, resulting in a periodic dimming. This periodic signature is used to detect the planet and to determine its size and its orbit. Three transits of a star, all with a consistent period, brightness change, and duration, provide a robust method of

TRANSIT (PLANETARY)

A planetary transit involves the passage of one celestial body in front of another, much larger-diameter celestial body. In solar system astronomy, one very important example is the transit of Venus across the face of the Sun, as seen by observers on Earth. Because of orbital mechanics, observers on Earth can witness only planetary transits of Mercury and Venus. There are about 13 transits of Mercury every century (100 years), but transits of Venus are much rarer events; in fact, only seven such events have occurred since the invention of the astronomical telescope. These transits took place in: 1631, 1639, 1761, 1769, 1874, 1882, and most recently on June 8, 2004. Any person who missed observing the 2004 transit should mark his (her) astronomical calendar because the next transit of Venus takes place on June 6, 2012.

Astronomers use contacts to characterize the principal events occurring during a transit. During one of the rare transits of Venus, for example, the event begins with contact I, which is the instant the planet's disk is externally tangent to the Sun. The entire disk of Venus is first seen at contact II, when the planet is internally tangent to the Sun. During the next several hours, Venus gradually traverses the solar disk at a relative angular rate of approximately 4 arcminutes per hour. At contact III, the planet reaches the opposite limb and is once again internally tangent to the Sun. The transit ends at contact IV, when the planet limb is externally tangent to the Sun. Contacts I and II define the phase of the transit called ingress, while astronomers refer to contacts III and IV as the egress phase or simply the egress.

From celestial mechanics, transits of Venus are only possible early December and early June when Venus's nodes pass across the Sun. If Venus reaches inferior conjunction at this time, a transit occurs. As you may have noticed from the list of the historic transits of Venus, these transits show a clear pattern of recurrence and take place at intervals of 8, 121.5, 8, and 105.5 years. The next pair of Venus transits will occur more than a century from now on December 11, 2117, and December 8, 2125.

So it is probably best to make plans to observe the upcoming June 6, 2012, transit of Venus. The entire 2012 transit (that is all four contacts) will be visible from northwestern North America, Hawaii, the western Pacific, northern Asia, Japan, Korea, eastern China, the Philippines, eastern Australia, and New Zealand. Unfortunately, no portion of the 2012 transit will be visible from Portugal or southern Spain, western Africa, and the southeastern two-thirds of South America. For the parts of the world not previously mentioned, the Sun will be setting or rising while the transit is in progress—so it will not be possible to observe the complete event (that is, all four contacts).

This is an artist's rendering of NASA's proposed Terrestrial Planet Finder (TPF) mission—a constellation of coorbiting telescopes with the mission to look for Earth-like planets that orbit nearby stars and to study the suitability of such planets as homes for any possible life. Scientists will use (ca. 2015) spectroscopic instruments on TPF to measure relative amounts of such gases as carbon dioxide, water vapor, ozone, and methane in the atmosphere of any detected terrestrial planets. These data, called biomarkers, will help exobiologists determine whether a planet is suitable for life—or perhaps whether life already exists there. (NASA/JPL)

detection and planet confirmation. The measured orbit of the planet and the known properties of the parent star are used to determine if each planet discovered is in the continuously habitable zone (CHZ), that is, at the distance from its parent star where liquid water could exist on the surface of the planet.

The *Kepler* spacecraft will hunt for planets using a specialized 3-foot- (1-meter-) diameter telescope called a photometer. This instrument can measure the small changes in brightness caused by the transits. By monitoring 100,000 stars similar to the Sun for four years following the *Kepler* spacecraft's launch, scientists expect to find many hundreds of terrestrial-type planets.

NASA's proposed Terrestrial Planet Finder (TPF) mission will consist of a suite of two complementary space observatories: a visible-light coronagraph and a midinfrared formation-flying interferometer. An interferometer consists of a collection of several (small) telescopes that function together to produce an image that is much sharper than would be possible with a single telescope. As currently planned, the Terrestrial Planet Finder coronagraph (TPF-C) should launch in 2014, and the four spacecraft making up the Terrestrial Planet Finder interferometer (TPF-I) by 2020. The combination will detect and characterize Earth-like planets around as many as 150 stars as many as up to 45 light-years away.

The science goals of the ambitious TPF mission include a survey of nearby stars in the search for terrestrial-size planets in the continuously habitable zone (CHZ). Scientists will then perform spectroscopy on the most promising candidate extrasolar planets, looking for atmospheric signatures that are characteristic of habitability or even life itself.

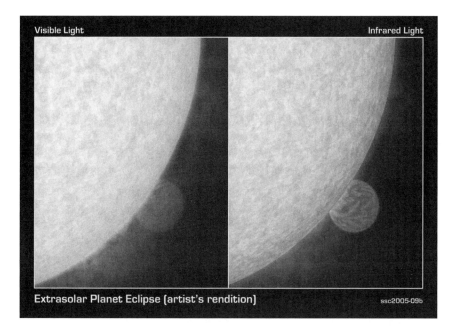

This artist's rendering shows how a fiery hot star and its nearby planetary companion might look close up if viewed in visible (left) and infrared (right) light. In visible light, a star shines brilliantly, overwhelming the little (visible) light that is reflected by the nearby planet. However, when viewed in the infrared portion of the electromagnetic spectrum, a star is much less blinding, and its companion planet perks up in a more readily observable thermal glow. A hot (5,700 K) yellow star, like the Sun, is brightest in the visible (yellow-light) wavelengths, while thermal radiation from a warm (300 K) planet peaks in the infrared portion of the EM spectrum. *(NASA/JPL–Caltech/R. Hurt [SSC])*

INFRARED ASTRONOMY

Infrared (IR) astronomy is the branch of modern astronomy that studies and analyzes infrared (IR) radiation from celestial objects. Most celestial objects emit some quantity of infrared radiation. However, when a star is not quite hot enough to shine in the visible portion of the electromagnetic spectrum, it emits the bulk of its energy in the infrared. IR astronomy, consequently, involves the study of relatively cool celestial objects, such as interstellar clouds of dust and gas (typically about -280°F [100 kelvins]) and stars with surface temperatures below temperatures below about 10,340°F (6,000 kelvins).

Many interstellar dust and gas molecules emit characteristic infrared signatures that astronomers use to study chemical processes occurring in interstellar space. This same interstellar dust also prevents astronomers from viewing visible light coming from the center of our Milky Way galaxy. However, IR radiation from the galactic nucleus is absorbed not as severely as radiation in the visible portion of the electromagnetic spectrum, and IR astronomy enables scientists to study the dense core of the Milky Way.

Infrared astronomy also allows astrophysicists to observe stars as they are being formed (they call these objects protostars) in giant clouds of dust and gas (called nebulae), long before their thermonuclear furnaces have ignited and they have "turned on" their visible emission.

Unfortunately, water and carbon dioxide in Earth's atmosphere absorb most of the interesting IR radiation arriving from celestial objects. Earth-based astronomers can use only a few narrow IR spectral bands or windows in observing the universe, and even these IR windows are distorted by "sky noise" (undesirable IR radiation from atmospheric molecules). With the arrival of the space age, however, astronomers have placed sophisticated IR telescopes (such as the *Spitzer Space Telescope*) in space, above the limiting and disturbing effects of Earth's atmosphere, and have produced comprehensive catalogs and maps of significant infrared sources in the observable universe.

The challenge of finding an Earth-size (terrestrial) planet orbiting even the closest stars can be compared to finding a tiny firefly next to a blazing searchlight when both are thousands of miles (km) away. Quite similarly, the infrared emissions of a parent star are a million times brighter than the infrared emissions of any companion planets that might orbit around it. Beyond the year 2020, data from the Terrestrial Planet Finder mission should allow astronomers to analyze the infrared emissions of extrasolar planets in star systems up to about 100 light-years away. They will then use these data to search for signs of atmospheric gases, such as carbon dioxide, water vapor, and ozone. Together with the temperature and radius of any detected planets, these atmospheric gas data will enable scientists to determine which extrasolar planets are potentially habitable or even if they may be inhabited by rudimentary forms of life.

✧ Summary of Extrasolar Planet Hunting Techniques

The quest for extrasolar planets is one of the most interesting areas within modern astronomy. But how can astronomers find planetary bodies around distant stars? In their search for extrasolar planets, scientists employ several important techniques. These techniques include: pulsar timing, Doppler spectroscopy, astrometry, transit photometry, and gravitational microlensing.

Astronomers accomplished the first widely accepted detection of extrasolar planets (namely, pulsar planets) in the early 1990s using the pulsar timing technique. Earth mass and even smaller planets orbiting a pulsar were detected by measuring the periodic variation in the pulse arrival time. However, the planets detected are orbiting a pulsar—a dead star, rather than a dwarf (main-sequence) star. What is encouraging, however, about the detection is that the planets were probably formed after the supernova that resulted in the pulsar, thereby demonstrating that planet formation is probably a common rather than a rare astronomical phenomenon.

Astronomers may use Doppler spectroscopy to detect the periodic velocity shift of the stellar spectrum caused by an orbiting giant planet. Scientists sometimes refer to this method as the radial velocity method. Using ground-based astronomical observatories, spectroscopists can measure Doppler shifts greater than about 9.8 feet per second (3 m/s) due to the reflex motion of the parent star. This measurement sensitivity corresponds to a minimum detectable planetary mass equivalent to approximately 30 Earth masses for a planet located at one astronomical unit. This method can be used for main-sequence stars of spectral types mid-F through M. Stars hotter and more massive than mid-F spectral class stars rotate faster, pulsate, are generally more active, and have less spectral structure, thus making measurement of their Doppler shifts much more difficult. As previously mentioned, using this technique, scientists have now successfully detected many large (Jupiter-like) extrasolar planets.

Scientists can also use astrometry to look for the periodic wobble that a planet induces in the position of its parent star. The minimum detectable planetary mass becomes smaller in inverse proportion to that planet's distance from the star. For a space-based astrometric instrument, such as NASA's planned *Space Interferometry Mission* (SIM)—a facility that will measure an angle as small as two micro-arcseconds—a minimum planet of about 6.6 Earth masses could be detected as it travels in a one-year orbit around a one solar-mass star that is 32.61 light-years (10 parsecs) from the Earth. The SIM PlanetQuest would also be capable of detecting a

0.4-Jupiter-mass planet in a four-year orbit around a star to a distance of 1,630 light-years (500 pc). From the ground, modern telescope facilities, such as the Keck Interferometer (in Hawaii) can measure angles as small as 20 micro-arcseconds, leading to a minimum detectable planetary mass in a one AU orbit of 66 Earth masses for a one solar-mass star at a distance of 32.6 light-years (10 pc). The limitations to this method are the distance to the star and variations in the position of the photometric center due to star spots. There are only 33 nonbinary solarlike (F, G, and K) main-sequence stars within 32.6 light-years (10 pc) of the Earth. The farthest planet from its star that can be detected by this technique is limited by the time needed to observe at least one orbital period.

Astronomers use the transit photometry technique to measure the periodic dimming of a star as caused by a planet passing in front of it along the line of sight from the observer. Stellar variability on the time scale of a transit limits the detectable size to about half that of Earth for a

SPACE INTERFEROMETRY MISSION

The *Space Inteferometry Mission (SIM)*, also called SIM PlanetQuest, is NASA's planned Origins Program mission now scheduled for launch in 2011. The *SIM PlanetQuest* spacecraft will be the first that has as its primary mission the performance of astrometry by long-baseline interferometry.

Astrometry involves the precise measurement of the positions and motions of objects in the sky. Free from the distortions and the noise of Earth's atmosphere, and with a maximum baseline of 32.8 feet (10 m), *SIM* will enable astronomers for the first time to measure the positions and motions of stars with micro-arcsecond accuracy. This represents a capability of detecting the small reflex motion of stars induced by orbiting planets a few times the mass of Earth.

SIM will conduct a detailed survey for planetary companions to stars in the solar neighborhood and will give scientists a more complete picture of the architecture of plan-etary systems around a representative sample of different stellar types. *SIM* will determine the positions and distances of stars throughout the Milky Way hundreds of times more accurately than any previous program and will create a stellar reference grid, providing a visible-light astronomical reference frame with unprec-edented precision. Against this reference frame, *SIM* will measure the internal dynamics of the Milky Way galaxy and the dynamics in the local group of galaxies, measure the photometric and astrometric effects of condensed dark matter in the galactic halo, and calibrate the brightness of several classes of astronomical "standard candles."

The Michelson Science Center (MSC) is responsible for developing and operating the Science Operations System (SOS) for *SIM*— including program solicitation, user interface and consultation, data infrastructure, and (jointly with the Jet Propulsion Laboratory [JPL]) science operations.

one AU orbit about a one solar-mass star; or, with four years of observing, transit photometry can detect Mars-size planets in Mercury-like orbits. Mercury-size planets can even be detected in the continuously habitable zone of K and M stars. Planets with orbital periods greater than two years are not readily detectable since their chance of being properly aligned along the line of sight to the star becomes very small.

Giant planets in inner orbits are detectable independent of the orbit alignment, based on the periodic modulation of their reflected light. The transit depth can be combined with the mass found from Doppler data to determine the density of the planet as scientists have done for the case of a star called HD209458b. Doppler spectroscopy and astrometry measurements can be used to search for any giant planets that might also be in the systems discovered using photometry. Doppler spectroscopy is based on the phenomenon called Doppler shift and the fact that a star does not remain completely stationary with respect to a planet orbiting around it. The parent star actually moves ever so slightly (in a small circle or ellipse) in response to the gravitational tug of a lower mass planetary companion. When observed by astronomers from Earth, these subtle movements cause a Doppler shift in the star's normal spectrum of light. If the star is being tagged slightly toward the observer, then its spectral light will appear slightly shifted toward the blue portion of the spectrum. Conversely, if the star is being slightly tugged away from Earth by its planetary companion, then its characteristic spectral light will appear slightly shifted toward the red portion of the spectrum. Since the orbital inclination must be close to 90 degrees to cause transits, there is very little uncertainty in the mass of any detected giant planet. Photometry represents the only practical method for finding Earth-size planets in the continuously habitable zone.

In the gravitational microlensing approach to extrasolar planet detection, scientists take advantage of an interesting physical phenomenon that is associated with Einstein's general theory of relativity—namely that gravity bends space. In May 1919, the British astronomer Sir Arthur Eddington (1882–1944) led a solar eclipse expedition to Principe Island (West Africa) to measure the gravitational deflection of a beam of starlight as it passed close to the Sun. His successful experiment provided early support for Einstein's newly introduced general relativity theory. Astrophysicists say that the presence of a very massive object (such as a star) warps the space-time continuum, causing a beam of light from a distant source to bend, as first demonstrated by Eddington in 1919 and by many other experimenters since then.

When a planet happens to pass in front of a parent star, the planet's gravity will actually behave like a lens and focus the rays of starlight. This causes a sharp but temporary increase in the parent star's brightness and an apparent change in the star's position when viewed by an observer along the line-of-sight. Consequently, astronomers sometimes find it convenient to use the effect of gravitational microlensing to detect (massive) objects that do not emit their own light. The parent star's light is focused (through gravitational microlensing) by gravity as a planet passes between the star and Earth.

✧ Biosignatures of Living Worlds

Exobiological exploration beyond this solar system is founded on the postulate that signatures of life (biosignatures) from a habitable and inhabited planet will be recognizable by remote sensing techniques that are conducted across interstellar distances. A biosignature is an object, substance, or pattern whose origin specifically requires a biological agent creating it. To be scientifically useful, a biosignature should also have a low probability of any nonbiological processes causing it.

The search for habitable planets around nearby stars is based on the further assumption that even the most basic life-forms on a suitable extrasolar planet will be global in extent. Exobiologists also assume that characteristic biosignatures, which are the evidence for life, from the habitable planet's surface or atmosphere will be readily recognizable in the spectrum of the planet's infrared light.

NASA's Terrestrial Planet Finder mission will use direct imaging detection and spectroscopic characterization techniques to detect the large-scale effects and signs of life on Earth-like planets around nearby stars. By analyzing the "colors" (wavelengths) of infrared radiation detected by TPF, astronomers and exobiologists hope to search for atmospheric gases such as carbon dioxide (CO_2), water vapor (H_2O), and ozone (O_3). Together with the surface temperature and size (diameter) of detected extrasolar planets, scientists can establish which of these extrasolar planets is habitable and perhaps even whether some are inhabited by basic forms of life.

Using this solar system and life on Earth as a reference, exobiologists suggest that the best candidates that merit closer investigation will be any found planets located in the continuously habitable zone (CHZ) around their parent star. Planets outside the CHZ are either too hot or too cold to support life as understood here on Earth. If a planet is too hot, all liquid water on the planet's surface eventually becomes vapor and then slowly escapes into space from the atmosphere. If the planet is too cold, all liquid water on the surface freezes. In the extremes, the former condition is called a runaway greenhouse effect, and the latter condition is called an ice catastrophe. Either of these environmental extremes would make even an Earth-sized extrasolar planet very inhospitable for life.

As a point of comparison, in humans' solar system, the continuously habitable zone starts beyond Venus (which is too hot) and ends before Mars (which is too cold and can no longer retain liquid water on its surface).

Scientists suggest that just existence of large amounts of oxygen (O_2) in a planet's atmosphere could be regarded as a strong biosignature. For example, in Earth's atmosphere, oxygen is a byproduct of photosynthesis—the fundamental process by which green plants and certain other organisms convert carbon dioxide and water into carbohydrates, using sunlight as the energy source. However, oxygen molecules do not linger long in a planet's atmosphere. Atmospheric oxygen experiences a process called oxidation, recombining with other types of molecules, so if they detected an extrasolar planet in the continuously habitable zone, scientists would be tempted to conclude that some type of life is keeping the supply of oxygen replenished on the planet.

However, scientists also recognize that there are nonbiological processes that could result in an oxygen-rich atmosphere on an abiotic planet, so conservative exobiologists do not regard the presence of oxygen in an extrasolar planet's atmosphere as an unambiguous biosignature. These cautious scientists would prefer to also detect ozone coexisting with nitrous oxide or methane before deciding whether an extrasolar planet is habitable. Once the Terrestrial Planet Finder mission provides evidence that a planet has oxygen, ozone, and methane in its atmosphere, for example, then the exobiologists consider such observational data convincing evidence not only that the candidate planet is habitable but also that it is inhabited.

NASA planners envision taking the results provided by the Terrestrial Planet Finder mission to shape the development of an even more sophisticated mission called *Life Finder*. Between 2020 and 2025, NASA would fly an array of large telescopes into space. These telescopes would combine infrared light from candidate extrasolar planets to create high-resolution spectra of their atmospheres. Exobiologists would then carefully examine these data for unambiguous indicators of biological activity, such as seasonal variations in the levels of methane and periodic changes in atmospheric chemistry. Of course, the scientists must remain ever mindful that life on Earth is their only known reference and that suspected biosignatures from candidate alien worlds may not appear exactly as the characteristic "life signs" produced by planet Earth.

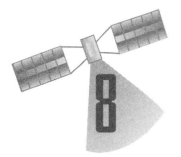

Search for Extraterrestrial Intelligence (SETI)

I n exobiology, the widely used acronym SETI means the search for extraterrestrial intelligence (SETI). The major goal of most publicly and privately funded SETI programs has been to listen for evidence of radio frequency (microwave) signals that are generated by intelligent extraterrestrial civilizations. A less frequently encountered companion acronym in the field of exobiology is CETI, which means communication with extraterrestrial intelligence. While the two terms appear comparable, they involve distinctly different approaches to interstellar contact and, therefore, could result in significantly different social consequences on a planetary scale. (See also chapter 11.)

SETI represents a more conservative, passive scientific approach to interstellar contact in which scientists on Earth patiently look for signs or listen for signals that may be indicative of the presence of intelligent alien beings elsewhere in the Milky Way galaxy. If a nonnatural, artifact signal is ever successfully received and decoded as part of some contemporary or future SETI effort, the people of Earth would be under no obligation to respond, and the sender of this signal would (in all likelihood) have no way of knowing that the message was intercepted by intelligent beings on a life-bearing planet called Earth that orbits around a main-sequence yellow star called Sol.

CETI involves a less conservative, more active technical approach to achieving interstellar contact. In the practice of CETI, terrestrial scientists intentionally send out specially prepared radio frequency signals or information-laden artifacts into the galaxy—announcing the presence of the human race to whomever or whatever out there is capable of receiving and interpreting these active attempts at interstellar communication. (See chapter 9.) In its most comprehensive form, CETI also implies that the people of Earth would respond to an alien signal, should one ever be received and successfully decoded.

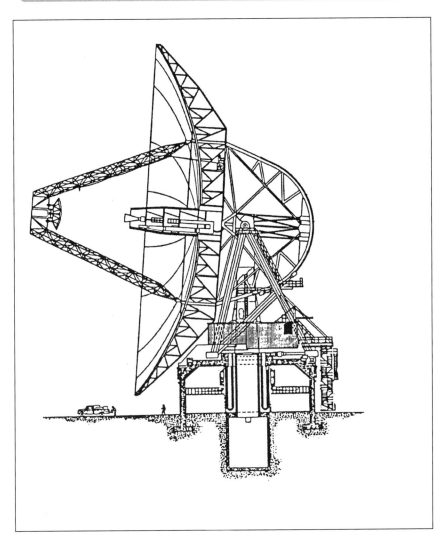

A large parabolic radio telescope, or "dish," as it is often called *(NASA)*

The large radio telescope is an important tool of modern astronomy that under certain operational conditions also supports both CETI and SETI activities. When used as a powerful radio frequency transmitter to beam a specially prepared signal to some interesting portion of the galaxy, the radio telescope becomes an important tool in the practice of CETI. When used as a very sensitive receiver to search for and collect faint "artifact" radio signals as may emerge from among the stars, the radio telescope becomes an important tool in the practice of SETI. The task for scientists, then, becomes one of skillfully separating the natural cosmic radio signals of interest to astronomers from any possible artificial (alien-produced artifact) signals of interest to exobiologists.

ANNAPOLIS MIDDLE SCHOOL
MEDIA CENTER

DISCOVERY OF THE PULSAR

The pulsar is a celestial object (thought to be a young, rapidly rotating neutron star) that emits radiation in the form of rapid pulses with a characteristic pulse period and duration. In August 1967, a then-graduate student Jocelyn Bell Burnell (1943–) and her academic advisor, the British astronomer and Nobel laureate Antony Hewish (1924–), detected the first pulsar. The unusual celestial object emitted radio waves in a pulsating rhythm. Because of the structure and repetition in the radio signal, their initial inclination was to consider the possibility that the repetitive radio signal was really from an intelligent extraterrestrial civilization. However, subsequent careful investigation and the discovery of another radio wave pulsar that December quickly dispelled the "little green man" signal hypothesis and showed the British scientists that the unusual signal was being emitted by a very interesting, newly discovered natural phenomenon—the pulsar.

Antony Hewish collaborated with Sir Martin Ryle (1918–84) in the development of the radio wave–based astrophysics. His efforts included the discovery of the pulsar—for which he shared the 1974 Nobel Prize in physics with Ryle. Although his graduate student Jocelyn Bell Burnell was actually the first person to notice the repetitive signals from this pulsar, the 1974 Nobel awards committee inexplicably overlooked her contributions. However, while giving his Nobel lecture, Hewish publicly acknowledged his student's observational contributions in the discovery of the first pulsar.

The detection of the first pulsar, a natural phenomenon of great interest in astronomy, was originally regarded (with a good degree of British humor) as a "little green man" signal. Some powerful radio telescopes, like the Arecibo Observatory in Puerto Rico (see chapter 9), are used in both capacities, namely in the role as a powerful radio transmitter and in the role as a supersensitive radio frequency receiver.

The social consequences of interstellar contact and the potential risks inherent in either SETI or CETI are discussed in chapter 11. This chapter concentrates on the technical and operational aspects of the more conservative "listen only" approach that is characteristic of modern SETI activities.

Humankind's search for intelligent life beyond Earth is an attempt to answer the important philosophical question: Is our species alone in the universe? The classic paper by Giuseppe Cocconi and Philip Morrison entitled "Searching for Interstellar Communications" (*Nature*, 1959) is often regarded as the start of modern SETI. With the arrival of the space age, the entire subject of extraterrestrial intelligence departed from the realm of science fiction and became treated—within many technical circles at least—as a scientifically respectable (although currently speculative) field of endeavor. Unfortunately, budget-axe-wielding politicians and visionless federal bureaucrats continue to find it fashionable and convenient to attack any attempt to commit even very modest quantities of federal research monies to the SETI effort. SETI scientists find it difficult to understand how research funding has been unquestionably approved by agencies of the federal government to investigate the mechanics of bovine flatulence but withheld from any serious attempt to pursue scientifically an answer to one of humankind's most enduring philosophical questions.

The current understanding of stellar formation leads scientists to postulate that planets are normal and frequent companions of most stars. As interstellar clouds of dust and gas condense to form stars, they appear to leave behind clumps of material that form into planets. Astronomers estimate that the Milky Way galaxy contains some 100 billion to 200

This artist's rendering depicts a young star encircled by full-sized planets and rings of dust beyond. These rings, or debris disks, arise when embryonic planets smash into each other. The sequence of events for one of these catastrophic planetary body collisions appears in the upper left inset. The illustration shows how planetary systems arise out of massive collisions between rocky bodies. NASA's *Spitzer Space Telescope* is collecting data, which indicates that these catastrophes continue to occur around stars when they are as old as 100 million years and even after the alien solar systems have developed full-sized planets. *(NASA/JPL–Caltech)*

billion stars. Starting in the 1990s, the detection of extrasolar planets by astronomers has validated this important assumption that planets are a normal byproduct of stellar evolution from clouds of dust and gas. (See chapter 7.) Consequently, there may be billions upon billions of planets in the galaxy—some, as yet an undetermined percentage of which, could be suitable for life.

Given an anticipated abundance of suitable planets, current theories on the origin and chemical evolution of life indicate that life probably is not unique to Earth but may be common and widespread throughout the galaxy. Furthermore, some scientists suggest that, once started, life on alien worlds will strive to evolve—leading in creatures with intelligence, curiosity, and even the technology to build the devices needed to transmit and receive electromagnetic signals across the interstellar void. For example, many intelligent alien civilizations (should such exist) might, like humans on Earth, radiate electromagnetic energy into space. This can happen unintentionally, as a result of planetwide radio-frequency communications networks; or intentionally through the deliberate beaming of structured radio signals out into the galaxy in the hope some other intelligent species can intercept and interpret these signals against the natural electromagnetic radiation background of space.

SETI observations may be performed using radio (and other) telescopes on Earth, in space, or even (someday) on the far side of the Moon. Each location has distinct advantages and disadvantages. Until recently, only very narrow portions of the electromagnetic spectrum have been examined for "artifact signals" (that is, those generated by intelligent alien civilizations). Humanmade radio and television signals—the kind that radio astronomers reject as clutter and interference—actually are similar to some of the type signals for which SETI researchers are hunting.

The sky is full of radio waves. In addition to the electromagnetic signals that human beings generate as part of their information-dependent technical civilization (for example, radio, TV, and radar), the sky also contains natural radio wave emissions from such celestial objects as the Sun, the planet Jupiter, radio galaxies, pulsars, and quasars. Even interstellar space is characterized by a constant, detectable radio-noise spectrum.

And just what would a radio-frequency signal from an intelligent extraterrestrial civilization look like? The accompanying figure presents a spectrogram that shows a simulated "artifact signal" from outside the solar system. This particular signal was sent by NASA's *Pioneer 10* spacecraft from beyond the orbit of Neptune and was received by a deep space network (DSN) radio telescope at Goldstone, California, using a 65,000-channel spectrum analyzer. The three signal components are quite visible above the always-present background radio noise. The center spike appearing in the figure has a transmitted signal power of approximately one watt, about half the power of a miniature Christmas tree light. SETI scientists are looking for a radio frequency signal that might appear this clearly but are also preparing their search equipment for one that may actually be more difficult to distinguish from the background radio wave noise (static). To search through a myriad of radio-frequency signals, SETI scientists have developed state-of-the-art spectrum analyzers that can sample millions of frequency channels simultaneously and can identify automatically candidate "artifact signals" for further observation and analysis.

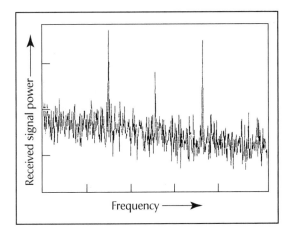

A simulated signal from an extraterrestrial civilization (using the *Pioneer 10* spacecraft to transmit an "artifact" signal from beyond the orbit of Neptune). *(NASA)*

In October 1992, NASA started a planned decade-long SETI program called the High Resolution Microwave Survey (HRMS). The main objective of HRMS was to search other solar systems for microwave signals, using radio telescopes at the National Science Foundation's Arecibo Observatory in Puerto Rico, NASA's Goldstone Deep Space Communications Complex in California, and other locations. Coupled with these telescopes were HRMS's, dedicated high-speed digital data-processing systems that contained off-the-shelf hardware and specially designed software.

The search proceeded in two different modes: a targeted search and an all-sky survey. The targeted search focused on about 1,000 nearby stars that resembled the Sun. In a somewhat less-sensitive search mode, the all-sky survey was planned to search the entire celestial sphere for unusual radio signals. However, severe budget constraints and refocused national research objectives resulted in the premature termination of NASA's HRMS program in 1993—after just one year of observation. Since then, while NASA has remained deeply interested in searching for life within our solar system, SETI projects no longer receive government funding.

Today, privately funded organizations and foundations (such as the SETI Institute in Mountain View, California—a nonprofit corporation that focuses on research and educational projects relating to the search for extraterrestrial life)—are conducting surveys of the heavens, hunting for radio signals from intelligent alien civilizations.

If an alien signal is ever detected and decoded, then the people of Earth would face another challenging question: Do we respond? For the present time, SETI scientists are content to listen passively for artifact signals that might arrive across the interstellar void.

✧ Radio Astronomy

Radio astronomy is the branch of astronomy that collects and evaluates radio signals from extraterrestrial sources. Radio astronomy is a relatively young scientific field, which started in the 1930s when the American radio engineer Karl Guthe Jansky (1905–50) detected the first extraterrestrial radio signals. Until Jansky's important discovery, astronomers had used only the visible portion of the electromagnetic spectrum to view the universe.

The detailed observation of cosmic radio sources is difficult, however, because these sources shed so little energy on Earth. But starting in the mid-1940s, with the pioneering work of the British astronomer Sir Alfred Charles Bernard Lovell (1913–   ) at the United Kingdom's Nuffield Radio Astronomy Laboratories at Jodrell Bank, the radio telescope has been used to discover some extraterrestrial radio sources so unusual that their very existence had not even been imagined or predicted by astrophysicists.

One of the strangest of these cosmic radio sources is the pulsar—a collapsed giant star that has become a neutron star and now emits pulsating radio signals as it rapidly spins. When the first pulsar was detected in 1967, it created quite a stir in the scientific community. Because of the regularity of its signal, scientists thought they had just detected the first interstellar signals from an intelligent alien civilization.

Another interesting celestial object is the quasar, or quasi-stellar radio source. Discovered in 1964, quasars now are considered to be entire galaxies in which a very small part (perhaps only a few light-days across) releases enormous amounts of energy—equivalent to the total annihilation of millions of stars. Quasars are the most distant known objects in the universe, and some of them are receding from Earth at more than 90 percent of the speed of light.

✧ Project Ozma

Project Ozma was the first attempt to detect interstellar radio signals from an intelligent extraterrestrial civilization. The American astronomer Frank Donald Drake (1930–   ) conducted this pioneering SETI experiment in 1960 at the National Radio Astronomy Observatory (NRAO) in Green Bank, West Virginia. Drake derived the name for his effort from the queen of the imaginary land of Oz since, in his own words, "Oz was a place very far away, difficult to reach, and populated by exotic beings."

He chose a radio frequency of 1,420 megahertz (MHz) for this initial search—the frequency of the 8.3-inch (21-cm) interstellar hydrogen line. Since this is a radio frequency at which most emerging technical civilizations would first use for narrow-bandwidth, high-sensitivity radio telescopes, Drake and other scientists had reasoned that this would also most likely be the frequency that more advanced alien civilizations would probably use in trying to signal emerging civilizations across the interstellar void. This line of thinking gave rise to the popular SETI term *water hole.*

In 1960, Drake aimed the approximately 85-foot- (26-m-) diameter Green Bank radio telescope at two sunlike stars, Tau Ceti and Epsilon Eridani, which are each about 11 light-years away. Patiently, he and members of his Project Ozma team listened for artifact radio wave signals that

WATER HOLE

Water hole is a SETI term used to describe a narrow portion of the electromagnetic spectrum that appears to be especially appropriate for interstellar communications between emerging and advanced civilizations. This band lies in the radio frequency (RF) part of the spectrum between 1,420 megahertz (MHz) frequency (8.3-inch [21.1-cm] wavelength) and 1,660 megahertz (MHz) frequency (7.1-inch [18-cm] wavelength).

Hydrogen (H) is abundant throughout interstellar space. When hydrogen experiences a "spin-flip" transition (due to an atomic collision), it emits a characteristic 1,420-megahertz frequency (or 8.3-inch [21.1-cm] wavelength) radio wave. Any intelligent species within the Milky Way galaxy that has developed radio astronomy will eventually detect these natural emissions. Similarly, there is another grouping of characteristic spectral emissions centered near the 1,660 megahertz frequency (7.1-inch [18-cm] wavelength) that are associated with hydroxyl (OH) radicals.

As scientists know from elementary chemistry, $H + OH = H_2O$. So there are, as suggested by SETI investigators, two radio-wave emission signposts that are associated with the dissociation products of water that "beckon all waterbased life to search for its kind at the age-old meeting place of all species: the water hole."

Is this high regard for the 1,420 to 1,660 megahertz frequency band reasonable, or is it simply a case of terrestrial chauvinism? Many exobiologists currently think that if life exists elsewhere in the universe, it will most likely be carbon based, and water is essential for carbon-based life, as known here on Earth. Furthermore, for purely technical reasons, when scientists scan all the decades of the electromagnetic spectrum in search for suitable frequency at which to send or receive interstellar messages, they ultimately arrive at the narrow microwave region between 1 and 10 gigahertz (GHz) as being the most suitable candidate region for conducting interstellar communication. The two characteristic emissions of dissociated water, namely 1,420 megahertz for H and 1,660 megahertz for OH, are situated right in the middle of this optimum communications band.

On the basis of this type of reasoning, the water hole has been favored by scientists engaged in SETI projects that involve the reception and analysis of radio signals. They generally consider that this portion of the electromagnetic spectrum represents a logical starting place for humans to listen for interstellar signals from other intelligent civilizations.

GREEN BANK TELESCOPE

On November 15, 1988, the original 300-foot (91.5-m) Green Bank radio telescope collapsed due to a sudden failure of a key structural element. This unexpected loss of the National Radio Astronomy Observatory (NRAO)'s major observing facility resulted in the construction of a replacement project, called the Robert C. Byrd Green Bank Telescope (GBT).

NRAO operates the Green Bank Telescope, which is the world's largest fully steerable single aperture antenna. In addition to the GBT, there are several other radio telescopes at the Green Bank site in West Virginia The GBT is generally described as a 328-foot (100-m) telescope, but the actual dimensions of the surface are 328 feet (100 m) by 361 feet (110 m). The overall structure of the GBT is a wheel-and-track design that allows the telescope to view the entire sky above 5 degrees elevation. The track, 210 feet (64 m) in diameter, is level to within a few thousandths of an inch (cm) to provide precise pointing of the structure while bearing 7,300 tons of moving mass. The GBT is of an unusual design. Unlike more conventional radio telescopes, which have a series of supports in the middle of the surface, the GBT's aperture is unblocked so that incoming radio frequency radiation meets the surface directly.

would indicate the presence of an intelligent alien civilization. But after more than 150 hours of monitoring, no evidence of strong radio signals from an intelligent extraterrestrial civilization was obtained. Project Ozma is generally considered as the first serious attempt to listen for interstellar radio signals from an intelligent alien civilization, making Drake's effort the birth of modern SETI.

Project Ozma led to Drake's formulation of a speculative, semiempirical mathematical expression, now popularly referred to as the Drake equation. As discussed more fully in a subsequent section, the Drake equation tries to estimate the number of intelligent alien civilizations that might now be capable of communicating with each other in the Milky Way galaxy. Among his many professional accomplishments in radio astronomy, Drake was a professor of astronomy at Cornell University (1964–84), served as the director of the Arecibo Observatory in Puerto Rico, and is currently emeritus professor of astronomy and astrophysics at the University of California at Santa Cruz.

✦ Project Cyclops

Project Cyclops was a study involving the proposed use of a very large array of dish antennas to perform in a detailed search of the radio-frequency spectrum (especially the 7.1-inch-to-8.3-inch [18-cm-to-21-cm] wavelength water-hole region) for artifact signals from intelligent alien civilizations. The engineering details of this SETI (search for extraterrestrial intelligence) configuration were derived in a special summer-institute design study that was sponsored by NASA at Stanford University in 1971. The stated object of the Project Cyclops study was to assess what would be required in hardware, human resources, time, and funding to mount a realistic effort, using present (or near-term future) state-of-the-art techniques, aimed at detecting the existence of intelligent life in other star systems.

Named for the one-eyed giants found in Greek mythology, the proposed Cyclops Project would use as its "eye" a large array of individually steerable 328-foot- (100-m-) diameter parabolic dish antennas. These Cyclops antennas would be arranged in a hexagonal matrix, so that each antenna unit was equidistant from all its neighbors. A 958-foot (292-m)

This artist's rendering provides a panoramic view of the Project Cyclops hexagonal array of 328-foot- (100-m-) diameter radio telescopes—as it might appear if the large facility were constructed and located on the far side of the Moon. In addition to terrestrial-noise-free radio astronomy, this large radio telescope complex would perform an extensive search for radio signals from intelligent, alien civilizations around other star systems. *(NASA)*

separation distance between antenna dish centers would help to avoid shadowing. In the Project Cyclops concept, an array of about 1,000 of these antennas would be used to collect and evaluate simultaneously radio signals falling on them from a target star system. The entire Cyclops array would function like a single giant radio antenna, some 11.6 square miles (30 km^2) to 23.2 square miles (60 km^2) in size. The study examined a Project Cyclops on Earth as well as on the far side of the Moon—an excellent radio astronomy location shielded from all stray terrestrial radio wave sources.

Project Cyclops can be regarded as one of the foundational studies in the search for extraterrestrial intelligence. Its results—based on the pioneering efforts of such individuals as Frank Drake (1930–   ), the late Philip Morrison (1915–2005), John Billingham (1930–   ), and the late Bernard Oliver (1916–95)—established the technical framework for subsequent SETI activities. Project Cyclops also reaffirmed the interstellar microwave window, the water hole, as perhaps the most suitable part of the electromagnetic spectrum for interstellar civilizations to communicate with each other.

✧ Drake Equation

The Drake equation is a probabilistic expression, proposed by the American astronomer Frank Drake in 1961. The expression is an interesting, though highly speculative, attempt to determine the number of advanced intelligent civilizations that might now exist in the Milky Way galaxy and might be communicating (via radio waves) across interstellar distances. A basic assumption in Drake's formulation is the principle of mediocrity—namely that conditions in the solar system and even on Earth are nothing particularly special but rather represent common conditions found elsewhere in the Milky Way galaxy.

Just where do scientists look among the billions of stars in the galaxy for possible interstellar radio messages or signals from extraterrestrial civilizations? That was one of the main questions addressed by the attendees of the Green Bank Conference on Extraterrestrial Intelligent Life held in November 1961 at the National Radio Astronomy Observatory (NRAO), Green Bank, West Virginia. One of the most significant and widely used results from this conference is the Drake equation, which represents the first credible attempt to quantify the search for extraterrestrial intelligence. This "equation" has also been called the Sagan-Drake equation and the Green Bank equation in the SETI literature.

Although more nearly a subjective statement of probabilities than a true scientific equality, the Drake equation attempts to express the number (N) of advanced intelligent civilizations that might be communicating across interstellar distances at the present time. The equation is generally expressed as:

$$N = R^* f_p n_e f_l f_i f_c L$$

where

N is the number of intelligent communicating civilizations in the galaxy at present,

R^* is the average rate of star formation in the Milky Way galaxy (stars/year),

f_p is the fraction of stars that have planetary companions,

n_e is the number of planets per planet-bearing star that have suitable ecospheres (that is, the environmental conditions necessary to support the chemical evolution of life),

f_l is the fraction of planets with suitable ecospheres on which life actually starts,

f_i is the fraction of planetary life starts that eventually evolve to intelligent life-forms,

f_c is the fraction of intelligent civilizations that attempt interstellar communication, and

L is the average lifetime (in years) of technically advanced civilizations.

An inspection of the Drake equation quickly reveals that the major terms cover many disciplines and that they vary in technical content from numbers that are somewhat quantifiable (such as R^*) to those that are completely subjective (such as L).

For example, astrophysics can provide a reasonably approximate value for R^*. Generally, the estimate for R^* used in SETI discussions is taken to fall between 1 and 20 stars per year.

The rate of planet formation in conjunction with stellar evolution is currently the subject of much interest and discussion in astrophysics. Do most stars have planets? If so, then the term f_p would have a value approaching unity. On the other hand, if planet formation is rare, then f_p approaches zero. Astronomers and astrophysicists now think that planets are a common occurrence in stellar-evolution processes. Furthermore, advanced extrasolar planet detection techniques (involving astrometry, adaptive optics, interferometry, spectrophotometry, high-precision radial velocity measurements, and several new "terrestrial planet hunter" spacecraft) will soon provide astronomers the ability to detect Earth-like planets (should such exist) around the Sun's nearest stellar neighbors. (See chapter 7.) Therefore, sometime within the next two decades, the search for extrasolar planets should provide scientists the number of direct observations needed to establish an accurate empirical value for f_p. In typical SETI discussions, f_p is now often assumed to fall in the range between 0.4 and 1.0. The value $f_p = 0.4$ represents a more pessimistic view, while the value $f_p = 1.0$ is taken as very optimistic.

Similarly, if planet formation is a normal part of stellar evolution, exobiologists must then ask how many of these planets are actually suitable for the evolution and maintenance of life? By taking $n_e = 1.0$, the scientists are suggesting that for each planet-bearing star system, there is at least one planet located in a suitable continuously habitable zone (CHZ), or ecosphere. This is, of course, what has occurred in humans' own solar system. Earth is comfortably situated in the continuously habitable zone, while Mars resides on the outer (colder) edge, and Venus is situated on the inner (warmer) edge.

The question of life-bearing "moons" around otherwise unsuitable planets was not directly addressed in the original Drake equation, but recent discussions about the existence of liquid-water oceans on several of Jupiter's moons, most notably Europa, and the possibility that such oceans might support life encourages exobiologists to consider a slight expansion of the meaning of the factor n_e. Perhaps n_e should now be taken to include the number of planets in a planet-bearing star system that lie within the habitable zone and also the number of major moons with potential life-supporting environments (liquid water, an atmosphere, etc.) around otherwise uninhabitable, Jovian-type planets in that same star system.

Scientists must next ask: Given conditions suitable for life, how frequently does it start? One important major assumption usually made by exobiologists (again invoking the principle of mediocrity) is that *wherever life can start, it will.* If they adhere to this assumption, then f_l equals unity. Similarly, most exobiologists like to assume that once life starts, it always strives toward the evolution of intelligence, making f_i equal to 1 (or extremely close to unity).

This brings the alien civilization-hunting scientists to an even more challenging question: What fraction of intelligent extraterrestrial civilizations develops the technical means and then want to communicate with other alien civilizations? All anyone can do here is make a very subjective guess, referencing human history. The pessimists take f_c to be 0.1 or less, while the optimists insist that all advanced civilizations desire to communicate and make $f_c = 1.0$.

At this point, it is appropriate to pause for a moment and consider the hypothetical case of an alien scientist in a distant star system (say about 50 light-years away) that must submit numerous proposals for very modest funding to continue a detailed search for intelligent-species-generated electromagnetic signals emanating from the Sun's region of the Milky Way. Unfortunately, the Grand Scientific Collective (the leading technical organization within that alien civilization) keeps rejecting this alien scientist's proposals stating that "such his/her/its proposed 'SETI' efforts are a waste of precious *zorbots* (alien unit of currency), which should be used for more worthwhile research projects." The alien scientist's super-sensitive radio receivers are turned off about a year before the first detectable television signals (leaking out from Earth) pass through that star system. Consequently, although this (hypothetical) alien civilization had developed the technology needed to justify use of a value of $f_c = 1$, that particular alien civilization did not display any serious inclination towards SETI—thereby making $f_c \approx 0$.

Extending the principle of mediocrity to the collective social behavior of intelligent alien beings (should they exist anywhere) is a highly speculative task. But, here is an important Drake equation–related question that should be considered. Is shortsightedness and a lack of strategic vision (especially among political leaders) a common shortcoming in otherwise intelligent creatures throughout the galaxy? If the principle of mediocrity is at work, the answer is, unfortunately, YES!

Finally, here on Earth, scientists must also speculate on how long an advanced technology civilization lasts. When they use Earth as a model, all they say is that (at a minimum) L is somewhere between 50 and 100 years. Truly high technology emerged on Earth only during the last century. Space travel, nuclear energy, computers, global telecommunications, and so on are now widely available on a planet that daily oscillates between the prospects of total destruction and a "golden age" of cultural and social maturity. Do most other evolving extraterrestrial civilizations follow a similar perilous pattern in which cultural maturity has to desperately race against new technologies that always threaten oblivion if they are unwisely used? Does the development of the technologies necessary for interstellar communication or perhaps interstellar travel also stimulate a self-destructive impulse in advanced civilizations, such that few (if any) survive? Or have many extraterrestrial civilizations learned to live with their evolving technologies, and do they now enjoy peaceful and prosperous "golden ages" that last for millennia to millions of years? In dealing with the Drake equation, the pessimists place very low values on L (perhaps a hundred or so years), while the optimists insist that L is several thousand to several million years in duration.

During any discussion concerning an appropriate value for L, it is interesting to recognize that space technology and nuclear technology also provide an intelligent species with very important tools for protecting their home planet from catastrophic destruction by an impacting "killer" asteroid or comet. (See chapter 2.) Although other solar systems will have cometary and asteroidal fluxes that are greater or less than those fluxes prevalent in humans' solar system, the threat of an extinction-level planetary collision might still be substantial. The arrival of high technology, therefore, also implies that intelligent aliens can overcome many natural hazards (including a catastrophic impact by an asteroid or comet)—

What is the lifetime of a typical advanced-technology civilization in the Milky Way galaxy? Do most (if not all) advanced civilizations end up destroying themselves with their own technologies before they learn to harness these powerful tools (including nuclear energy) and travel to other star systems? Depicted here is the atmospheric detonation of an 11-kiloton-yield nuclear device called FITZEAU, which was exploded by the United States at the Nevada Test Site on September 14, 1957. *(The U.S. Department of Energy/Nevada Operations Office)*

thereby extending the overall lifetime of the planetary civilization and increasing the value of L that scientists should use in the Drake equation.

Going back now to the Drake equation, some "representative" values are inserted. Taking $R^* = 10$ stars/year, $f_p = 0.5$ (thereby excluding multiple-star systems), $n_e = 1$ (based on humans' solar system as a common model), $f_l = 1$ (invoking the principle of mediocrity), $f_i = 1$ (again invoking the principle of mediocrity), and $f_c = 0.2$ (assuming that most advanced civilizations are introverts or have no desire for space travel), then the Drake equation yields: $N \approx L$. This particular result implies that the number of communicative extraterrestrial civilizations in the galaxy at present is approximately equal to the average lifetime (in Earth years) of such alien civilizations.

It is also instructive to take these "results" one step further. If N is about 10 million (a very optimistic Drake equation output), then the average distance between intelligent, communicating civilizations in the Milky Way galaxy is approximately 100 light-years. If N is 100,000, then these extraterrestrial civilizations on the average would be about 1,000 light-years apart. But if there were only 1,000 intelligent alien civilizations existing today, then they would typically be some 10,000 light-years apart. So, even if the Milky Way galaxy does contain a few such alien civilizations, they may be just too far apart to achieve communications within the total lifetimes of their respective civilizations. For example, at a distance of 10,000 light-years, it would take 20,000 years just to start an interstellar dialogue.

✧ Fermi Paradox—"Where Are They?"

A paradox is an apparently contradictory statement that may nevertheless be true. According to the lore of physics, the famous Fermi paradox arose one evening in 1943 during a social gathering at Los Alamos, New Mexico, when the brilliant Italian-American physicist and Nobel laureate Enrico Fermi (1901–54) asked the penetrating question: "Where are *they*?" "Where are who?" his startled companions replied. "Why, the extraterrestrials," responded the Nobel prize–winning physicist. At the time, Fermi and the other scientists in the group were involved in the top-secret Manhattan Project—the American effort to build the world's first atomic bomb.

Fermi's line of reasoning that led to this famous paradox has helped to form the basis of much modern thinking with respect to SETI. It can be summarized as follows: The Milky Way galaxy is some 13 to 15 billion (10^9) years old and contains perhaps 100 to 200 billion stars. If just one advanced civilization had arisen in this period of time and attained the technology necessary to travel between the stars, that advanced civilization could have diffused through, or traveled across, the entire galaxy within 50 million to 100 million years—leaping from star to star, starting up other civilizations, and spreading intelligent life everywhere.

But as scientists on Earth search the heavens, they do not see a galaxy teeming

The Italian–American physicist and Nobel laureate Enrico Fermi (1901–54). *(The U.S. Department of Energy/Argonne National Laboratory)*

with intelligent life, nor do they have any technically credible evidence of visitations or contact with alien civilizations, so they logically conclude that probably no such technically superior extraterrestrial civilization has ever arisen in the approximately 15-billion-year history of the galaxy. Here arises the great paradox: Although scientists might expect to see signs of a universe filled with intelligent life (on the basis of statistics and the number of possible "life sites"—given the existence of 100 billion to 200 billion stars in just this galaxy alone), they have collected no credible evidence of such. Are humans, then, really alone? If, on the other hand, we are not alone—where are *they?*

Many attempts have been made to respond to this very profound question. The "pessimists" reply that the reason human beings have not seen any signs of intelligent extraterrestrial civilizations is because humans really are alone in the galaxy and perhaps in the entire universe. One alternative sometimes suggested is that the human race is actually the galaxy's first technically advanced beings to rise to the level of space travel. If so, then perhaps it is humans' cosmic destiny to be the first species to sweep through the galaxy spreading intelligent life.

The "optimists," on the other hand, hypothesize that intelligent life exists out there somewhere and offer a variety of possible reasons why scientists on Earth have not yet "seen" signs of these alien civilizations. This section of the book discusses just a few of the many suggested reasons. First, perhaps intelligent alien civilizations really do not want anything to do with the human race. As an emerging planetary civilization, humans may be perceived as just too belligerent or too intellectually backward to bother with. Another possibility is that the communications technologies available in humans' planetary civilization simply lie below the minimum communications horizon of advanced alien civilizations. Other optimists suggest that not every intelligent civilization has the desire

ANCIENT ASTRONAUT THEORY

The ancient astronaut theory is a contemporary hypothesis that Earth was visited in the past by intelligent alien beings. Although *unproven* in the context of traditional scientific investigation, this hypothesis has given rise to many popular stories, books, and motion pictures. The ancient astronaut theory is built around unfounded speculations and extrapolations that seek to link such phenomena as the legends from ancient civilizations concerning superhuman creatures, unresolved mysteries of the past, and unidentified flying objects (UFOs) with extraterrestrial sources, such as visits by alien space travelers.

LABORATORY HYPOTHESIS

The laboratory hypothesis is a variation of the zoo hypothesis response to the Fermi paradox. This particular hypothesis stipulates that the reason human beings cannot detect or interact with technically advanced extraterrestrial civilizations in the Milky Way galaxy is because the intelligent aliens have set the solar system up as a "perfect" laboratory. The (hypothesized) alien scientists and researchers want to observe and study humans but do not want us to be aware of or influenced by their observations.

to travel between the stars or that maybe the intelligent alien species do not even desire to communicate by means of electromagnetic signals. Yet another response to the Fermi paradox is that *we* are actually *they*—the descendants of ancient star travelers who visited Earth millions of years ago when an alien wave of exploration and expansion passed through this part of the galaxy.

Another group of optimists might respond to Fermi's question by suggesting that intelligent aliens are out there right now but are keeping a safe distance. The aliens prefer to watch the human race either mature as a planetary civilization or else destroy itself. A subset of this response is called the extraterrestrial zoo hypothesis. This response hypothesis suggests that humans and Earth are being kept as a "zoo" or wildlife preserve by advanced alien zookeepers who have elected to monitor our activities but not be detected themselves. There is a terrestrial model for this hypothesis: Professional naturalists and animal care experts often recommend that a "perfect zoo" here on Earth would be one in which the animals being kept have no direct contact with their keepers and do not even realize that they are in captivity.

Finally, other people respond to the Fermi paradox by saying that the wave of cosmic expansion has not yet reached this section of the galaxy—so humans should keep looking. Within this response group are those who boldly declare that the alien visitors are now among us!

✧ Some Speculations about the Nature of Extraterrestrial Civilizations

According to some scientists, intelligent extraterrestrial (ET) civilizations in the galaxy might be conveniently characterized as falling into one of several basic levels of societal development or technology use. In SETI activities, one of the most widely used scales is the three types of ET civilizations that were introduced in 1964 by the Russian astronomer Nikolai Semenovich Kardashev (1932–   ). While examining the issue of interstellar information transmission by extraterrestrial civilizations, Kardashev postulated that there were just three basic types of technologically developed civilizations in the galaxy. He structured his proposed characterization scheme on the basis of how much total energy each type of ET civilization was able to harness and manipulate. He further suggested that the more energy a particular alien civilization controlled, the easier it would be for them to communicate across vast interstellar distances.

A Kardashev Type I civilization represents a planetary civilization similar to the technology level on Earth in the late 20th century. A typical planetary civilization would command the use of somewhere between 10^{12} and 10^{16} watts (J/s) of energy. Here, the upper limit of energy available is the total amount of radiant energy being intercepted by Earth (or an Earth-like, extrasolar planet in the continuously habitable zone) as the planet travels in orbit around the parent star.

With respect to Earth, scientists define the solar constant as the total amount of the Sun's radiant energy that normally crosses perpendicular to a unit area at the top of the planet's atmosphere. At one astronomical unit from the Sun, scientists have measured the value of the solar constant and it is approximately 127 watts/ft^2 (1,371W/m^2). Of course, an alien civilization might live on a planet that orbits a little closer to a less radiant and slightly cooler K-spectral class or M-spectral class parent star, so the total amount of energy available to a Type I alien civilization could vary somewhat from the suggested maximum value of 10^{16} watts (J/s).

A Kardashev Type II civilization would engage in feats of planetary engineering, emerge from its native planet through advances in space technology, and extend its resource base throughout the local star system. The eventual upper limit of a Type II civilization could be taken as the creation of a Dyson sphere. A Dyson sphere is a postulated shell-like cluster of habitats and structures placed entirely around a star by an advanced interplanetary civilization to intercept and to use basically all the radiant energy from the parent star. What the British-American physicist Freeman J. Dyson (1923–   ) suggested in 1960 was that an advanced extraterrestrial civilization might eventually develop the space technologies that were necessary to rearrange the raw materials of all the planets in its solar system, creating a more efficient composite ecosphere around the parent star. Using the radiant energy output of the Sun as the reference, a Kardashev Type II civilization would command between 10^{26} and 10^{27} watts (J/s) of energy. Once this size solar-system civilization is achieved, the search for additional resources and the pressures of continued growth could encourage the alien society to pursue interstellar migrations. Initiation of the process of interstellar migration would mark the start of a Kardashev Type III extraterrestrial civilization.

At maturity, a Type III civilization would be capable of harnessing the material and energy resources of an entire galaxy (typically containing some 10^{11} to 10^{12} stars). Energy resources on the order of 10^{37} to 10^{38} watts (J/s) or more would be involved.

Command of energy resources is an important figure of merit in comparing (hypothetical) extraterrestrial civilizations. Within Kardashev's scheme, a Type II civilization controls about 10^{12} times the energy resources of a Type I civilization, and a Type III civilization controls approximately 10^{12} times as much energy as a Type II civilization.

What else can scientists speculate about such civilizations? Again, starting with Earth as a model (at present, the one and only "scientific" data point), scientists can reasonably postulate that a Type I civilization would probably exhibit the following characteristics: (1) an understanding of the laws of physics; (2) a planetary society, including a global communication network and interwoven food and materials resource networks; (3) intentional or unintentional emission of electromagnetic radiations (especially radio frequency) into the galaxy; (4) the development of space technology and rocket propulsion-based interplanetary travel—the vital tools necessary to leave the home planet; (5) (possibly) the development of nuclear energy technology, both power supplies and weapons; and (6) (possibly) a desire to search for and communicate with other intelligent life-forms beyond the home planet.

Many uncertainties, of course, are present in such characterizations. For example, given the development of the space technology, will the planetary civilization decide to create a solar-system civilization? Do the planet's inhabitants develop a long-range planning perspective that supports the eventual creation of artificial habitats and structures throughout their star system? Or do the majority of Type I civilizations end up destroying themselves with their own advanced technologies before they can emerge from a planetary civilization into a more stable Type II solar-system civilization? Does the exploration imperative encourage intelligent alien creatures to go out from their comfortable planetary niche into an initially hostile but resource-rich solar system? If this "cosmic birthing" does not occur frequently, perhaps the Milky Way galaxy is indeed populated with intelligent life but at a level of stagnant planetary (Type I) civilizations that have neither the technology nor the societal motivation to create an extraterrestrial civilization. Such introvert Type I civilizations may not even to try to communicate with any other intelligent life-forms across interstellar distances.

Assuming that an extraterrestrial civilization does, however, emerge from its native planet and create an interplanetary society, several additional characteristics could become evident. The construction of space habitats and structures (leading ultimately to a Dyson sphere around the parent star) would portray feats of planetary engineering. Something as large as a Dyson sphere could possibly be detected by telltale thermal infrared emissions, as the parent star's radiant energy (predominantly in the visible spectrum) was intercepted and converted to other more useful forms of energy and as the residual energy (as determined by the universal laws of thermodynamics) was rejected to space as waste heat at perhaps 80°F (300K).

Type II civilizations might also decide to search in earnest for other forms of intelligent life beyond their solar system. Alien scientists in a Type II civilization would probably use portions of the electromagnetic spectrum (radio frequency and perhaps X-rays or gamma rays) as information carriers between the stars. Remembering that Type II civilizations would control 10^{12} times as much energy as Type I civilizations, such techniques as electromagnetic beacons or feats of astroengineering that yielded characteristic X-ray or gamma-ray signatures could lie well within their technical capabilities. Assuming their understanding of the physical universe is far more sophisticated than ours, Type II civilizations might also use gravity waves or other physical phenomena that are perhaps unknown at present by terrestrial scientists. In this case, the advanced alien civilization might now be sending such "exotic signals" through humans' solar system, but no equipment on Earth is currently capable of receiving and analyzing their efforts to communicate across vast interstellar distances. One currently difficult-to-detect signal would be a modulated beam of neutrinos. The field of neutrino astronomy here on Earth is in its infancy.

Type II civilizations might also decide to make initial attempts at interstellar matter transfer. Fully autonomous robotic explorers would be sent forth on one-way scouting missions to nearby stars. Even if the mode of propulsion involved devices that achieved only a small fraction of the speed of light, Type II societies should have developed the much longer-term strategic planning perspective necessary to support such sophisticated, expensive, and lengthy missions. The Type II civilization might also use a form of directed panspermia (the intentional diffusion of spores or molecular precursors through space). The alien civilization might employ robotic interstellar probes to disperse microscopically encoded viruses through the interstellar void, hoping that if such "seeds of life" found a suitable ecosphere in some neighboring or distant star system, they would initiate the chain of life perhaps leading ultimately to the replication (suitably tempered by local ecological conditions) of intelligent life itself.

Finally, as the Dyson sphere was eventually completed, some of the inhabitants of this (hypothetical) Type II civilization might respond to a cosmic wanderlust and initiate the first "intelligently crewed" interstellar missions. Complex space habitats could become space arks and carry portions of the advanced alien civilization to neighboring star systems.

Such scenarios are intriguing, but exobiologists also ask: What is the lifetime of a Type II civilization? It would appear from an extrapolation of contemporary terrestrial engineering practices that perhaps a minimum of 500 to 1,000 years would be required for an advanced interplanetary civilization to complete a Dyson sphere.

Throughout the entire galaxy, if just one Type II civilization embarks on a successful interstellar migration program, then—at least in principle—that alien race would eventually (in perhaps 10^8 to 10^9 years) sweep through the galaxy in a leapfrogging wave of exploration, establishing a Type III civilization in its wake.

This Type III civilization could eventually control the energy and material resources of up to 10^{12} stars—or the entire Milky Way galaxy. Communication or matter transfer would be accomplished by techniques that can now only politely be called "exotic." Perhaps very precisely manipulated beams of neutrinos or (hypothesized) faster-than-light particles (such as tachyons) would serve as the standard information carriers for this galactic society. Or, they might use tunneling through black holes as their transportation network. Perhaps, they might develop some kind of thought-transference or telepathic skills that permitted efficient and instantaneous communications across the vast regions of interstellar space. In any event, a Type III civilization should be readily evident since it would be galactic in extent and easily recognizable by its incredible feats of astroengineering and technical "magic."

In all likelihood, the Milky Way galaxy at present does not contain a Type III civilization, or else the solar system is being ignored—that is, intentionally being kept isolated—perhaps as a game preserve or zoo, as some scientists have speculated. Then again, the solar system may be simply one of the very last regions to be "filled in." These are some of the more popular speculations associated with the Fermi paradox.

There is also another important perspective: If the human race is really alone or is the most advanced civilization in the galaxy, then the men and women of Earth now stand at the technological threshold of creating the galaxy's first Type II civilization. Should the human race succeed in that task, our descendents would have the opportunity of becoming the first interplanetary (Type II) civilization to travel into interstellar space, founding a Type III civilization within the Milky Way galaxy.

Messages to the Stars: Arecibo Message, Pioneer Plaques, Voyager Records

While the human race started to leak its presence out into the galaxy unintentionally in the form of radio and (later) television signals early in the 20th century, there have also been three deliberate attempts to send messages to alien civilizations that may reside among the stars. The first attempt involved the broadcast of a powerful radio message, known as the Arecibo Interstellar Message. The other two deliberate attempts at interstellar communication involved artifacts, two special plaques and two records, placed aboard four NASA spacecraft (*Pioneer 10* and *11* and *Voyager 1* and *2*), whose missions to the outer planets ultimately placed the far-traveling robots on interstellar trajectories. This chapter describes these efforts.

✧ Arecibo Interstellar Message

To help inaugurate the powerful radio/radar telescope of the Arecibo Observatory in the tropical jungles of Puerto Rico, an interstellar message of friendship was beamed to the fringes of the Milky Way galaxy. On November 16, 1974, this interstellar radio signal was transmitted toward the Great Cluster in Hercules (Messier 13 or M13, for short), which lies about 25,000 light-years away from Earth. The globular cluster M13 contains about 300,000 stars within a radius of approximately 18 light-years.

This transmission, often called the Arecibo Interstellar Message, was made at the 2,380-megahertz (MHz) radio frequency with a 10 hertz (Hz) bandwidth. The average effective radiated power was 3×10^{12} watts (3 terawatts [TW]) in the direction of transmission. The signal is considered to be the strongest radio signal yet beamed out into space by our planetary civilization. Perhaps 25,000 years from now, a radio telescope operated by

ARECIBO OBSERVATORY

The Arecibo Observatory is the world's largest radio/radar telescope, with a 1,000-foot- (305-m-) diameter dish. It is located in a large, bowl-shaped natural depression in the tropical jungle of Puerto Rico. This huge facility is the main observing instrument of the National Astron-omy and Ionosphere Center (NAIC), a national center for radio and radar astronomy and ionospheric physics that is operated by Cornell University under contract with the National Science Foundation. The observatory operates on

(continues)

The Arecibo Observatory is the world's largest radio/radar telescope. The facility, located in a large, bowl-shaped natural depression in the tropical jungle of Puerto Rico, has a diameter of 1,000 feet (305 m). *(NASA)*

(continued)

a continuous basis, 24 hours a day every day, providing observing time and logistic support to visiting scientists.

When the giant telescope operates as a radio-wave receiver, it can listen for signals from celestial objects at the farthest reaches of the universe. As a radar transmitter/receiver, it assists astronomers and planetary scientists by bouncing signals off the Moon, off nearby planets and their satellites, off asteroids, and even off layers of Earth's ionosphere.

The Arecibo Observatory has made many contributions to astronomy and astrophysics. In 1965, the facility (operating as a radar transmitter/receiver) determined that the rotation rate of the planet Mercury was 59 days rather than the previously estimated value of 88 days. In 1974, the facility (operating as a radio-wave receiver) supported the discovery of the first binary pulsar system. This discovery led to an important confirmation of Albert Einstein's theory of general relativity and earned the American physicists Russell A. Hulse (1950–) and Joseph H. Taylor, Jr., (1941–) the 1993 Nobel Prize in physics. In the early 1990s, astronomers used the facility to discover extra-solar planets in orbit around the rapidly rotating pulsar B1257+12.

In May 2000, astronomers used the Arecibo Observatory as a radar transmitter/receiver to collect the first-ever radar images of a main-belt asteroid named 216 Kleopatra. Kleopatra is a large, dog-bone-shaped minor planet about 135 miles (217 km) long and 58.4 miles (94 km) wide. Discovered in 1880, the exact shape of Kleopatra was unknown until early this century. Astronomers used the telescope to bounce radar signals off Kleopatra. Then, with sophisticated computer analysis techniques, the scientists decoded the echoes, transformed them into images, and assembled a computer model of the asteroid's shape. This activity was made possible because the Arecibo radio telescope underwent major upgrades in the 1990s—improvements that dramatically improved its sensitivity and made feasible the radar imaging of more distant objects in the solar system.

The Arecibo Observatory is also uniquely suited to search for signals from extraterrestrial life, by focusing on thousands of star systems in the 1,000 MHz to 3,000 MHz range. To date, no confirmed SETI signals have been found.

members of an intelligent alien civilization somewhere in the M13 cluster will receive and decode this interesting signal. If they do, they will learn that intelligent life had evolved here on Earth.

The Arecibo Interstellar Message of 1974 consisted of 1,679 consecutive characters. It was written in a binary format—that is, only two different characters were used. In binary notation, the two different characters are denoted as "0" and "1." In the actual transmission, each character was represented by one of two specific radio frequencies, and the message was transmitted by shifting the frequency of the Arecibo Observatory's radio transmitter between these two radio frequencies in accordance with the plan of the message.

The message itself was constructed by the staff of the National Astronomy and Ionosphere Center (NAIC). It can be decoded by breaking up the message into 73 consecutive groups of 23 characters each and then arranging these groups in sequence one under the other. The numbers 73 and 23 are prime numbers. Their use should facilitate the discovery by any alien civilization receiving the message that the above format is the right way to interpret the message. The figure on page 182 shows the decoded message: The first character transmitted (or received) is located in the upper right-hand corner.

This message describes some of the characteristics of terrestrial life that the scientific staff at the National Astronomy and Ionosphere Center felt would be of particular interest and technical relevance to an extraterrestrial civilization. The NAIC staff interpretation of the interstellar message is as follows.

The Arecibo message begins with a "lesson" that describes the number system being used. This number system is the binary system, where numbers are written in powers of 2 rather than of 10, as in the decimal system used in everyday life. NAIC staff scientists believe that the binary system is one of the simplest number systems and is particularly easy to code in a simple message. Written across the top of the message (from right to left)

This artist's rendering, entitled "Call," represents SETI alien style. The interesting painting shows an alien scientist operating a large radio transmitter/receiver telescope array on a planet in a distant solar system. The alien is sending radio signals out into the Milky Way galaxy in an attempt to make contact with another intelligent species. The intelligent being is also using the facility to listen patiently for any radio signals that may arrive from other intelligent civilizations. (© *Pat Rawlings, the artist; used here with permission*)

Decoded form of the Arecibo Message of 1974. (Frank D. Drake and the staff of the National Astronomy and Ionosphere Center, operated by Cornell University under contract with the National Science Foundation)

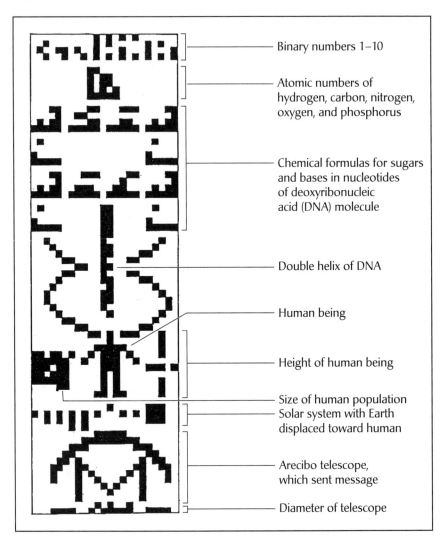

Binary numbers 1–10

Atomic numbers of hydrogen, carbon, nitrogen, oxygen, and phosphorus

Chemical formulas for sugars and bases in nucleotides of deoxyribonucleic acid (DNA) molecule

Double helix of DNA

Human being

Height of human being

Size of human population
Solar system with Earth displaced toward human

Arecibo telescope, which sent message

Diameter of telescope

are the numbers 1 through 10 in binary notation. Each number is marked with a *number label*—that is, a single character, which denotes the start of a number.

The next block of information sent in the message occurs just below the numbers. It is recognizable as five numbers. From right to left these numbers are: 1, 6, 7, 8, and 15. This otherwise unlikely sequence of numbers should eventually be interpreted as the atomic numbers of the elements hydrogen, carbon, nitrogen, oxygen, and phosphorus.

Next in the message are 12 groups on lines 12 through 30 that are similar groups of five numbers. Each of these groups represents the chemical formula of a molecule or radical. The numbers from right to left in each case provide the number of atoms of hydrogen, carbon, nitrogen, oxygen, and phosphorus, respectively, that are present in the molecule or radical.

Since the limitations of the message did not permit a description of the physical structure of the radicals and molecules, the simple chemical formulas do not define in all cases the precise identity of the radical or molecule. However, these structures are arranged as they are organized within the macromolecule described in the message. Intelligent alien organic chemists somewhere in the M13 cluster should eventually be able to arrive at a unique solution for the molecular structures being described in the message.

The most specific of these structures, and perhaps the one that should point the way to interpreting the others correctly, is the molecular structure that appears four times on lines 17 through 20 and lines 27 through 30. This is a structure containing one phosphorus atom and four oxygen atoms, the well-known phosphate group. The outer structures on lines 12 through 15 and lines 22 through 25 give the formula for a sugar molecule, deoxyribose. The two sugar molecules on lines 12 through 15 have between them two structures: the chemical formulas for thymine (left structure) and adenine (right structure). Similarly, the molecules between the sugar molecules on lines 22 through 25 are: guanine (on the left) and cytosine (on the right).

The macromolecule or overall chemical structure is that of deoxyribonucleic acid (DNA). The DNA molecule contains the genetic information that controls the form, living processes, and behavior of all terrestrial life. This structure is actually wound as a double helix, as depicted in lines 32 through 46 of the message. The complexity and degree of development of intelligent life on Earth is described by the number of characters in the genetic code, that is, by the number of adenine-thymine and guanine-cytosine combinations in the DNA molecule. The fact that there are some 4 billion such pairs in human DNA is illustrated in the message by the number given in the center of the double helix between lines 27 and 43. Note that the number label is used here to establish this portion of the message as a number and to show where the number begins.

The double helix leads to the "head" in a crude sketch of a human being. The scientists who composed the message hoped that this would indicate connections among the DNA molecule, the size of the helix, and the presence of an "intelligent" creature. To the right of the sketch of a human being is a line that extends from the head to the feet of the "message human." This line is accompanied by the number 14. This portion of the message is intended to convey the fact that the "creature" drawn is 14 units of length in size. The only possible unit of length associated with the message is the wavelength of the transmission, namely about five inches (12.6 cm). This makes the creature in the message some five feet nine inches (176 cm) tall. To the left of the human being is a number, four billion. This number represents the approximate human population on planet Earth when the message was transmitted.

Below the sketch of the human being is a representation of humans's solar system. The Sun is at the right, followed by nine planets with some coarse representation of relative sizes. The third planet, Earth, is displaced to indicate that there is something special about it. In fact, it is displaced toward the drawing of the human being, who is centered on it. Hopefully, an extraterrestrial scientist in pondering this message will recognize that Earth is the home of the intelligent creatures that sent it.

Below the solar system and centered on the third planet is an image of a telescope. The concept of *telescope* is described by showing a device that directs rays to a point. The mathematical curve leading to such a diversion

WHAT DO YOU SAY TO A LITTLE GREEN MAN?

Suppose there is an intelligent extraterrestrial species out there, say 75 light-years away, who is not only capable of receiving our radio messages but also wants to communicate. If humans really are successful at SETI, just what will we talk about with this alien race? Remember, radio waves travel at the speed of light—so it will take 75 years for a message to travel each way in this hypothetical interstellar communication. There are going to be some very long pauses in this extraterrestrial "phone call." In fact, the first interstellar conversation with all the excitement it will cause on Earth might start something like this:

EARTH: This is Earth calling. Is anyone there?

(—very long pause of about 150 years—)

THEM: (*possible response number one*)

Yes, we are here. Who (or what) are you?

or

THEM: (*possible response number two*)

Yes, someone is here. What do you want?

or finally

THEM: (*possible response number three*)

Yes, someone is here. Who and *where* are you?

Well, just how should the people of Earth respond to any of these hypothetical replies? Furthermore, who should speak for Earth?

Maybe we could just send out some basic data and news about the human race—much like the Arecibo interstellar message sent in 1974. But that is almost like a holiday greeting card with a little, nonpersonalized printed newsletter tucked inside. People of Earth really have to choose the items selected for transmission carefully to guarantee the durability of the information. For example, if scientists decide to send news about their latest technical achievements, after the passage of a century and a half or so, humans could become very embarrassed by the "backward" civilization that these data appear to represent. Just think of the technical changes that have taken place on Earth in the last century and a half. Perhaps human beings might decide to send of paths is crudely indicated. The telescope is not upside down but rather "up" with respect to the symbol for the planet Earth.

At the very end of the message, the size of the telescope is indicated. Here, it is both the size of the largest radio telescope on Earth and also the size of the telescope that sent the message (namely, the radio telescope at the Arecibo Observatory). It is shown as 2,430 wavelengths across, or roughly 1,000 feet (305 m). No one, of course, expects an alien civilization to have the same unit system that we use here on Earth—but physical quantities, such as the wavelength of transmission, provide a common reference frame.

This interstellar message was transmitted at a rate of 10 characters per second, and it took 169 seconds to transmit the entire information package. It is interesting to realize that just one minute after completion

information about the current world situation or a collection of welcoming speeches from various terrestrial politicians and leaders. But we may quickly lose our extraterrestrial audience with such boring, domestic trivia. Even the hottest news items of today will in all likelihood become rather insignificant in the overall context of planetary history.

So what do the people of Earth say to the "little green men" (LGMs) who have been so patiently waiting for radio messages from other star systems? One scientist feels that humans could talk about mathematics, physics, and astronomy. The long pauses between exchanges might give subsequent generations of humans something to look forward to—as they await the next message from the stars. Another scientist has suggested that Earth send samples of human-created music, and, if advanced radio frequency or optical frequency communications techniques permit, even images of great works of art. Perhaps the creation and appreciation of beauty might serve as a common basis of communication between civilizations throughout the galaxy.

Along these lines, the first communication sequence between beings from Earth and an extraterrestrial civilization could boil down to something like:

EARTH: Hello! Is anyone there?

(—very long pause of about 150 years—)

THEM: Yes, we're here. Who are we talking to?

(—very long pause again of about 150 years—)

EARTH: We are intelligent beings, called humans, from a place called Earth; listen to some of our civilization's most beautiful music. . . .

(—very long pause again of about 150 years—)

THEM: You've got a wrong number. (Click!)

Seriously, however, before the human race attempts to reach out and touch someone or something across interstellar space, people, acting in unison as a planetary society, must think about how we might respond to an interstellar message that invites Earth's civilization to "chat." Who should speak for the people of planet Earth? What information should they provide about Earth? And what questions should they ask about the alien civilization?

of transmission, the interstellar greetings passed the orbit of Mars. After 35 minutes, the message passed the orbit of Jupiter, and after 71 minutes, it silently crossed the orbit of Saturn. Some 5 hours and 20 minutes after transmission, the message passed the orbit of Pluto, leaving the solar system and entering "interstellar space." The Arecibo interstellar message will be detectable by telescopes anywhere in the Milky Way galaxy of approximately the same size and capability as the terrestrial facility that sent it.

✧ Interstellar Journeys of the Pioneer 10 and 11 Spacecraft

The *Pioneer 10* and *11* spacecraft, as their names imply, are true deep-space explorers—the first humanmade objects to navigate the main asteroid belt, the first spacecraft to encounter Jupiter and its fierce radiation belts, the first to encounter Saturn, and the first spacecraft to leave the solar system. These spacecraft also investigated magnetic fields, cosmic rays, the solar wind, and the interplanetary dust concentrations as they flew through interplanetary space.

The *Pioneer 10* spacecraft was launched from Cape Canaveral Air Force Station, Florida, by an Atlas-Centaur rocket on March 2, 1972. It

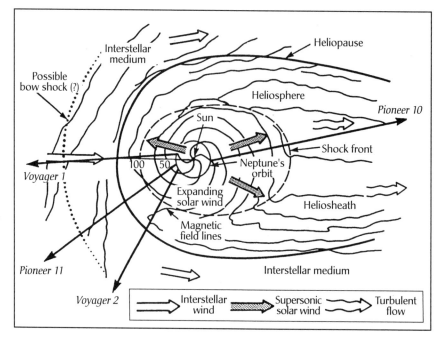

Paths of the *Pioneer 10* and *11* spacecraft, as well as the *Voyager 1* and *2* spacecraft, through the heliosphere and into the interstellar medium. (NASA)

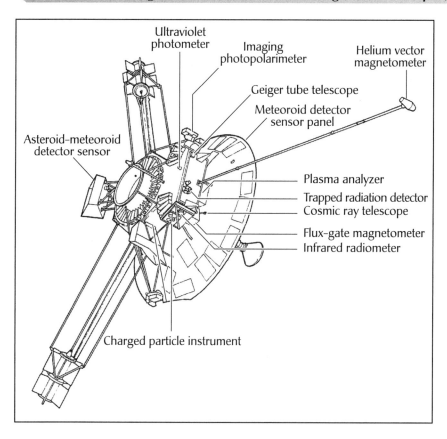

Ultraviolet photometer

Imaging photopolarimeter

Helium vector magnetometer

Geiger tube telescope

Meteoroid detector sensor panel

Asteroid-meteoroid detector sensor

Plasma analyzer

Trapped radiation detector

Cosmic ray telescope

Flux-gate magnetometer

Infrared radiometer

Charged particle instrument

The *Pioneer 10* and *11* spacecraft with their complement of scientific instruments. Each spacecraft's electric power was provided by a long-lived radioisotope thermoelectric generator (RTG). *(NASA)*

became the first spacecraft to cross the main asteroid belt and the first to make close-range observations of the Jovian system. Sweeping past Jupiter on December 3, 1973 (its closest approach to the giant planet), it discovered no solid surface under the thick layer of clouds enveloping the giant planet—an indication that Jupiter is a liquid hydrogen planet. *Pioneer 10* also explored the giant Jovian magnetosphere, made close-up pictures of the intriguing Red Spot, and observed at relatively close range the Galilean satellites Io, Europa, Ganymede, and Callisto. When *Pioneer 10* flew past Jupiter, it acquired sufficient kinetic energy to carry it completely out of the solar system.

Departing Jupiter, *Pioneer 10* continued to map the *heliosphere* (the Sun's giant magnetic bubble, or field, drawn out from it by the action of the solar wind). Then, on June 13, 1983, *Pioneer 10* crossed the orbit of Neptune, the major planet farthest out from the Sun. The historic date marked the first passage of a humanmade object beyond the major planet boundary of the solar system. Once across this solar system boundary, *Pioneer 10* continued to measure the extent of the heliosphere as the spacecraft started its travels into interstellar space. Along with its sister ship (*Pioneer 11*), the *Pioneer 10* spacecraft helped scientists investigate the deep space environment.

The *Pioneer 10* spacecraft is heading generally toward the red star, Aldebaran. The robot spacecraft is more than 68 light-years away from Aldebaran, and the journey will require about 2 million years to complete. Budgetary constraints forced NASA to terminate routine tracking and project data-processing operations for *Pioneer 10* on March 31, 1997. However, occasional tracking of *Pioneer 10* continued beyond that date. The last successful data acquisitions from *Pioneer 10* by NASA's Deep Space Network (DSN) occurred in 2002 on March 3 (30 years after launch) and again on April 27. The spacecraft signal was last detected on January 23, 2003, after an uplink message was transmitted to turn off the remaining operational experiment, the Geiger Tube Telescope. However, no downlink data signal was achieved, and by early February 2003 no signal at all was detected. NASA personnel concluded that the spacecraft's radioisotope thermoelectric generator (RTG) unit, which supplied electric power, had finally fallen below the level needed to operate the onboard transmitter. Consequently, no further attempts were made to communicate with *Pioneer 10*.

The *Pioneer 11* spacecraft was launched on April 5, 1973, and swept by Jupiter at an encounter distance of only 26,725 miles (43,000 km) on December 2, 1974. The spacecraft provided additional detailed data and pictures of Jupiter and its moons, including the first views of Jupiter's polar regions. Then, on September 1, 1979, *Pioneer 11* flew by Saturn, demonstrating a safe flight path through the rings for the more sophisticated *Voyager 1* and *2* spacecraft to follow. *Pioneer 11* (by then officially renamed *Pioneer Saturn*) provided the first close-up observations of Saturn, its rings, satellites, magnetic field, radiation belts, and atmosphere. The space robot found no solid surface on Saturn, but discovered at least one additional satellite and ring. After rushing past Saturn, *Pioneer 11* also headed out of the solar system toward the distant stars.

The *Pioneer 11* spacecraft operated on a backup transmitter since launch. Instrument power sharing began in February 1985 due to declining RTG power output. Science operations and daily telemetry ceased on September 30, 1995, when the RTG power level became insufficient to operate any of the spacecraft's instruments. All contact with *Pioneer 11* ceased at the end of 1995. At that time, the spacecraft was 44.7 astronomical units (AU) away from the Sun and traveling through interstellar space at a speed of about 2.5 AU per year.

Both *Pioneer* spacecraft carry a special message (called the Pioneer plaque) for any intelligent alien civilization that might find them wandering through the interstellar void millions of years from now. This message is an illustration, engraved on an anodized aluminum plaque. The plaque depicts the location of Earth and the solar system, a man and a woman,

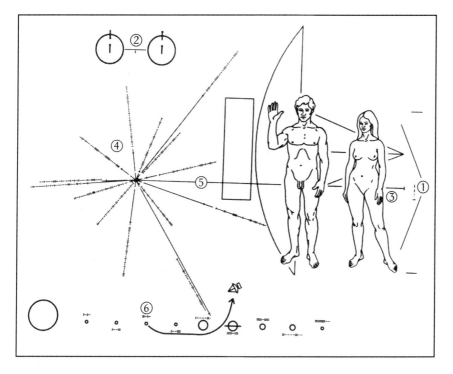

Annotated *Pioneer 10* and *11* plaque. Numbers are superimposed on the original plaque to facilitate discussion: (1) shows height of the woman compared to the *Pioneer* spacecraft; (2) is a schematic of the hyperfine transition of neutral atomic hydrogen; (3) represents the number 8 in binary form; the radial pattern (4) locates the solar system within the Milky Way galaxy; the solid bars indicate distance, with the long horizontal bar (5) with no binary notation on it representing the distance from the Sun to the galactic center, while the shorter solid bars denote directions and distances to 14 pulsars from the Sun; and (6) is a diagram of humans' solar system. *(NASA)*

and other points of science and astrophysics that should be decipherable by a technically intelligent civilization.

The plaque is intended to show any intelligent alien civilization that might detect and intercept either *Pioneer* spacecraft millions of years from now when the spacecraft was launched, from where it was launched, and by what type of intelligent beings it was built. The plaque's design is engraved into a gold-anodized aluminum plate, 6 inches (15.2 cm) by 9 inches (22.9 cm). The plate is approximately 0.05 inches (0.127 cm) thick. Engineers attached the plaque to the *Pioneer* spacecraft's antenna support struts in a position that helps shield it from erosion by interstellar dust.

The previous figure shows an annotated version of the Pioneer plaque. The numbers (1 to 6) have been intentionally superimposed on the plaque to assist in the discussion of its message. At the far right, the bracketing bars (1) show the height of the woman compared to the *Pioneer* spacecraft. The drawing at the top left of the plaque (2) is a schematic of the hyperfine transition of neutral atomic hydrogen used here as a universal "yardstick" that provides a basic unit of both time and space (length) throughout the Milky Way galaxy. This figure illustrates a reverse in the direction of the spin of the electron in a hydrogen atom. The transition depicted emits a characteristic radio wave with an approximately 8.3-inch (21-cm) wavelength. Therefore, by providing this drawing, the people of Earth are telling any technically knowledgeable alien civilization finding it that they have chosen 8.3 inches (21 cm) as a basic length in the message. While extraterrestrial civilizations will certainly have different names and defining dimensions for their basic system of physical units, the wavelength size that is associated with the hydrogen radio-wave emission will still be the same throughout the galaxy. Science and commonly observable physical phenomena represent a general galactic language—at least for starters.

The horizontal and vertical ticks (3) represent the number 8 in binary form. Hopefully, the alien beings pondering over this plaque will eventually realize that the hydrogen wavelength (8.3 inch [21 cm]) multiplied by the binary number representing 8 (indicated alongside the woman's silhouette) describes her overall height—namely, 8 × 8.3 inches = 66 inches (8 x 21 cm = 168 cm), or approximately five and one half-feet tall. Both human figures are intended to represent the intelligent beings that built the *Pioneer* spacecraft. The man's hand is raised as a gesture of goodwill. These human silhouettes were carefully selected and drawn to maintain ethnic neutrality. Furthermore, no attempt was made to explain terrestrial "sex" to an alien culture—that is, the plaque makes no specific effort to explain the potentially mysterious differences between the man and woman who are depicted.

The radial pattern (4) should help alien scientists locate the solar system within the Milky Way galaxy. The solid bars indicate distance, with the long horizontal bar (5) with no binary notation on it representing the distance from the Sun to the galactic center, while the shorter solid bars denote directions and distances to 14 pulsars from the Sun. The binary digits following these pulsar lines represent the periods of the pulsars. From the basic time unit that was established by the use of the hydrogen-atom transition, an intelligent alien civilization should be able to deduce that all times indicated are about 0.1 second—the typical period of pulsars. Since pulsar periods appear to be slowing down at well-defined rates, the pulsars serve as a form of galactic clock. Alien scientists should be able to search their astrophysical records and identify the star system from which the *Pioneer* spacecraft originated and approximately when it was launched, even if each spacecraft is not found for hundreds of millions of years. Consequently, through the use of this pulsar map, NASA's engineers and scientists have attempted to locate Earth, both in galactic space and in time.

As a further aid to identifying the *Pioneer 10*'s (or *Pioneer 11*'s) origin, a diagram of the solar system (6) is also included on the plaque. The binary digits accompanying each planet indicate the relative distance of that planet from the Sun. The spacecraft's trajectory is shown starting from the third planet (Earth), which has been offset slightly above the others. As a final clue to the terrestrial origin of the *Pioneer* spacecraft, its antenna is depicted pointing back to Earth.

This message was designed for NASA by Frank Drake (1930–   ) and the late Carl Sagan (1934–96), and Linda Salzman Sagan (1940–   ) prepared the artwork.

✧ Voyager Interstellar Mission (VIM)

As the influence of the Sun's magnetic field and solar wind grow weaker, both *Voyager* robot spacecraft eventually will pass out of the heliosphere and into the interstellar medium. Through NASA's Voyager Interstellar Mission (VIM) (which began officially on January 1, 1990), the *Voyager 1* and *2* spacecraft will continue to be tracked on their outward journey.

The two major objectives of the VIM are an investigation of the interplanetary and interstellar media and a characterization of the interaction between the two, and a continuation of the successful Voyager program of ultraviolet astronomy. During the VIM, the spacecraft will search for the heliopause (the outermost extent of the solar wind, beyond which lies interstellar space). Scientists hope that at least one *Voyager* spacecraft will still be functioning when it penetrates the heliopause and will provide them with the first true sampling of the interstellar environment. Barring a catastrophic failure on board either *Voyager* spacecraft, their nuclear power systems should provide useful levels of electric power until at least 2015.

Each *Voyager* spacecraft has a mass of 1,815 pounds (825 kg) and carries a complement of scientific instruments to investigate the outer planets and their many moons and intriguing ring systems. These instruments that are provided electric power by a long-lived nuclear system called a radioisotope thermoelectric generator (RTG) recorded spectacular close-up images of the giant outer planets and their interesting moon systems, explored complex ring systems, and measured properties of the interplanetary medium.

Once every 176 years, the giant outer planets—Jupiter, Saturn, Uranus, and Neptune—align themselves in such a pattern that a spacecraft launched from Earth to Jupiter at just the right time might be able to visit the other three planets on the same mission, using a technique called gravity assist. NASA space scientists named this multiple giant-planet encounter mission the Grand Tour and took advantage of a unique celestial alignment opportunity in 1977 by launching two sophisticated spacecraft, called *Voyager 1* and *2*.

The *Voyager 2* spacecraft lifted off from Cape Canaveral, Florida, on August 20, 1977, on board a Titan-Centaur rocket. (NASA called the first *Voyager* spacecraft launched *Voyager 2* because the second *Voyager* spacecraft to be launched eventually would overtake it and become *Voyager 1*.) *Voyager 1* was launched on September 5, 1977. This spacecraft followed the same trajectory as its twin (*Voyager 2*) and overtook its sister ship just after entering the asteroid belt in mid-December 1977.

Voyager 1 made its closest approach to Jupiter on March 5, 1979, and then used Jupiter's gravity to swing itself to Saturn. On November 12, 1980, *Voyager 1* successfully encountered the Saturn system and then was flung up out of the ecliptic plane on an interstellar trajectory. The *Voyager 2* spacecraft encountered the Jupiter system on July 9, 1979 (closest approach), and then used the gravity assist technique to follow *Voyager 1* to Saturn. On August 25, 1981, *Voyager 2* encountered Saturn and then went

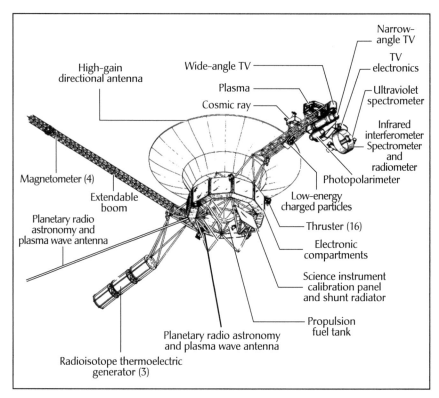

The *Voyager 1* and *2* spacecraft and their complement of sophisticated scientific instruments. *(NASA)*

on to encounter both Uranus (January 24, 1986) and Neptune (August 25, 1989) successfully. Space scientists consider the end of *Voyager 2*'s encounter of the Neptunian system as the end of a truly extraordinary epoch in planetary exploration.

During the first 12 years since they were launched from Cape Canaveral, these incredible robot spacecraft contributed more to the understanding of the giant outer planets of the solar system than was accomplished in more than three millennia of Earth-based observations. Following its encounter with the Neptunian system, *Voyager 2* also was placed on an interstellar trajectory and (like its *Voyager 1* twin) now continues to travel outward from the Sun.

Since both *Voyager* spacecraft would eventually journey beyond the solar system, their designers placed a special interstellar message on each in the hope that perhaps millions of years from now some intelligent alien race will find either spacecraft drifting quietly through the interstellar void. If they are able to decipher the instructions for using this record, they will learn about the contemporary terrestrial civilization and the men and women who sent *Voyager* on its stellar journey.

The gold-plated copper record entitled "The Sounds of Earth" is now being carried into interstellar space on board both the *Voyager 1* and *2* spacecraft. *(NASA)*

The Voyager interstellar message is a phonograph record called "The Sounds of Earth." Electronically imprinted on it are words, photographs, music, and illustrations that will tell an extraterrestrial civilization about planet Earth. Included are greetings in more than 50 different languages, music from various cultures and periods, and a variety of natural terrestrial sounds such as the wind, the surf, and different animals. The Voyager record also includes a special message from former President Jimmy Carter (1924–   ). The late Carl Sagan (1934–96) described in detail the full content of this phonograph message to the stars in his book *Murmurs of Earth*.

Each record is made of copper with gold plating and is encased in an aluminum shield that also carries instructions on how to play it. The following figure shows the set of instructions that accompany the Voyager record. In the upper left is a drawing of the phonograph record and the stylus carried with it. Written around it in binary notation is the correct

The set of instructions to any alien civilization that might find the *Voyager 1* or *2* spacecraft and that explains how to operate the Voyager record and from where the spacecraft and message came. *(NASA)*

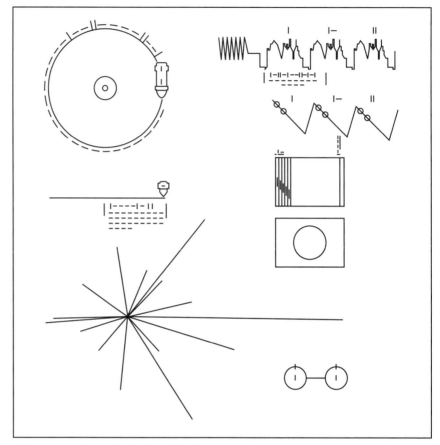

time for one rotation of the record, 3.6 seconds. Here, the time units are 0.70 billionths of a second, the time period associated with a fundamental transition of the hydrogen atom. The drawing further indicates that the record should be played from the outside in. Below this drawing is a side view of the record and the stylus, with a binary number giving the time needed to play one side of the record (approximately one hour).

The information provided in the upper-right portion of the instructions is intended to show how pictures (images) are to be constructed from the recorded signals. The upper-right drawing illustrates the typical waveform that occurs at the start of a picture. Picture lines 1, 2, and 3 are given in binary numbers and the duration of one of the picture "lines" is also noted (about 8 milliseconds). The drawing immediately below shows how these lines are to be drawn vertically, with a staggered interlace to give the correct picture rendition. Immediately below this is a drawing of an entire picture raster, showing that there are 512 vertical lines in a complete picture. Then, immediately below this is a replica of the first picture on the record. This should allow extraterrestrial recipients to verify that they have properly decoded the terrestrial pictures. A circle was selected for this first picture to guarantee that any aliens who find the message use the correct aspect ratio in picture reconstruction.

Finally, the drawing at the bottom of the protective aluminum shield is that of the same pulsar map drawn on the *Pioneer 10* and *11* plaques. The map shows the location of the solar system with respect to 14 pulsars, whose precise periods are also given. The small drawing with two circles in the lower-right-hand corner is a representation of the hydrogen atom in its two lowest states, with a connecting line and digit 1. This indicates that the time interval associated with the transition from one state to the other is to be used as the fundamental time scale, both for the times given on the protective aluminum shield and in the decoded pictures.

Spreading Life Around: Designing Interstellar Probes and Starships

This chapter encourages each reader to take a giant intellectual leap and consider future technologies as might be developed by the human race and made available to perform space exploration several centuries, or even a millennium, from now. This mental exercise includes some really over-the-horizon technologies that involve interstellar robot probes, self-replicating systems that could wander through the entire galaxy, and even starships that are capable of carrying intelligent machines and their human companions. The chapter focuses on technologies that populate a technical horizon, which would allow the human race to spread life deliberately into outer space. The first sequence of such activities would be spreading life throughout humans' home solar system; the next major step would involve sending a wave of life and consciousness into a portion, and then perhaps across all, of the Milky Way galaxy. Space exploration visionaries and advocates sometimes use the term *greening the galaxy* to describe the diffusion of human beings, their technology, and their culture through interstellar space, first to neighboring star systems and then eventually across the entire galaxy.

The chapter begins with a discussion of the first lunar bases and Martian settlements that will give life beyond Earth a firm foothold in the solar system. A future generation of space-faring humans (real Selenites and Martians) in partnership with incredibly sophisticated machines could transform neighboring worlds into suitable habitats by using amazing feats of planetary engineering. As humans' solar-system civilization approaches maturity, it will be characterized in the ultimate by the creation of a Dyson sphere. At that future point, the descendents of today's global population may decide to venture into interstellar space—first with robot probes and then with sophisticated robots called self-replicating systems, crewed starships, or perhaps some combination of both technical options. In the third millennium, this robot-human partnership in space exploration will make the universe both a destination and a destiny for our species.

✦ Lunar Bases and Settlements

There are many factors (some favorable and some unfavorable) and physical resource assessments that will dramatically influence and shift any lunar base–development scenario that will be suggested in 2007. Recognizing such limitations in contemporary technical projections, this section provides a generalized overview of what might take place during the remainder of this century if human space-exploration activities include the development of a permanent lunar base.

When human beings return to the Moon, it will not be for a brief moment of scientific inquiry as occurred in NASA's Apollo Project but rather as permanent inhabitants of a new world. They will build bases from which to explore the lunar surface completely, establish science and technology laboratories that take advantage of the special properties of the lunar environment, and harvest the Moon's resources (including

Just three days away from Earth by rocket-propelled space travel, the Moon is a good place to test the hardware and operations for the first human expedition to Mars. This artist's concept shows a simulated Mars mission, including the landing of a lunar environment-adapted Mars excursion vehicle, which could test many relevant Mars expedition systems and technologies. *(NASA/JSC; artist, Pat Rawlings)*

the suspected deposits of lunar ice in the polar regions) in support of humanity's extraterrestrial expansion.

A lunar base is a permanently inhabited complex on the surface of the Moon. In the first permanent lunar-base camp, a team of from 10 up to perhaps 100 lunar workers will set about the task of fully investigating the Moon. The word *permanent* here means that human beings will always occupy the facility, but individuals probably will serve tours of from one to three years before returning to Earth. Some workers at the base will enjoy being on another world. Some will begin to experience isolation-related psychological problems, similar to the difficulties often experienced by members of scientific teams, who "winter-over" in Antarctic research stations. Still other workers will experience injuries or even fatal accidents while working at or around the lunar base.

For the most part, however, the pioneering lunar-base inhabitants will take advantage of the Moon as a science-in-space platform and perform the fundamental engineering studies needed to confirm and define the specific roles that the Moon will play in the full development of space for the remainder of this century and in centuries beyond. For example, the discovery of frozen volatiles (including water) in the perpetually frozen recesses of the Moon's polar regions could change lunar-base logistics strategies and accelerate development of a large lunar settlement of as many as 10,000 or more inhabitants.

Many lunar-base applications have been proposed. Some of these concepts include: (1) a lunar scientific laboratory complex; (2) a lunar industrial complex to support space-based manufacturing; (3) an astrophysical observatory for solar system and deep space surveillance; (4) a fueling station for orbital transfer vehicles that travel through cislunar space; and (5) a training site and assembly point for the first human expedition to Mars.

Social and political scientists suggest that a permanent lunar base could also become the site of innovative political, social, and cultural developments—essentially rejuvenating our concept of who we are as intelligent beings and boldly demonstrating our ability to apply advanced technology beneficially in support of the positive aspects of human destiny. Another interesting suggestion for a permanent lunar base is its use as a field operations center for the rapid response portion of a planetary defense system that protects Earth from threatening asteroids or comets.

As lunar activities expand, the original lunar base could grow into an early settlement of about 1,000 more-or-less permanent residents. Then, as the lunar industrial complex develops further and lunar raw materials, food, and manufactured products start to support space commerce throughout cislunar space, the lunar settlement itself will expand to a population of about 10,000. At that point, the original settlement might

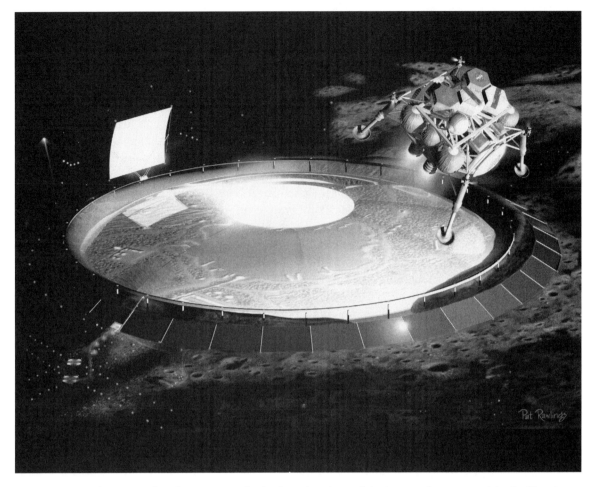

Scientists currently suspect that the permanently shadowed regions of the lunar poles may contain significant deposits of water-ice. If this turns out to be true, harvesting this precious resource would be an important part of any permanent human occupancy of the Moon. This artist's concept shows a solar-powered base in a crater at the Moon's south pole, harvesting lunar water-ice and producing rocket propellant (hydrogen and oxygen) for lunar spacecraft, like the one illustrated. In this depiction, the base's inhabitants are circulating some of the water that they have harvested through the dome's cells to provide additional shielding against space radiation. *(NASA/ JSC; artist, Pat Rawlings)*

spawn several new settlements—each taking advantage of some special location or resource deposit elsewhere on the lunar surface.

In the next century, this collection of permanent human settlements on the Moon could continue to grow, reaching a combined population of about 500,000 persons and attaining a social and economic "critical mass" that supports true self-sufficiency from Earth. This moment of self-sufficiency for the lunar civilization will also be a very historic moment in human history. With the rise of a self-sufficient, autonomous lunar civilization, future generations will have a choice of worlds on which to live and to prosper, for from that time on, the human race will exist in two distinct and separate "biological niches"—people will be either *terran* or *nonterran* (that is, extraterrestrial).

The vast majority of lunar-base development studies include the use of the Moon as a platform from which to conduct science in space. Scientific facilities on the Moon will take advantage of its unique environment to support platforms for astronomical, solar, and space science (plasma) observations. The unique environmental characteristics of the lunar surface include low gravity (one-sixth that of the Earth), high vacuum, seismic stability, low temperatures (especially in permanently shadowed polar regions), and a low radio-noise environment on the Moon's farside.

Astronomy from the lunar surface offers the distinct advantages of a low radio-noise environment and a stable platform in a low-gravity environment. The farside of the Moon is permanently shielded from direct terrestrial radio emissions. As future radio-telescope designs approach their ultimate (theoretical) performance limits, this uniquely quiet lunar environment may be the only location in all cislunar space where sensitive radio-wave-detection instruments can be used to full advantage, both in radio astronomy and in our search for extraterrestrial intelligence (SETI). In fact, radio astronomy, including extensive SETI efforts, may represent one of the main "lunar industries" later in this century. In one sense, SETI performed by lunar-based scientists could be viewed as "extraterrestrials" searching for other extraterrestrials.

The Moon also provides a solid, seismically stable, low-gravity, high-vacuum platform for conducting precise interferometric and astrometric observations. For example, the availability of ultrahigh-resolution (micro-arcsecond) optical, infrared, and radio observatories will allow astronomers to search carefully for Earth-like extrasolar planets that encircle nearby stars, to a distance of perhaps several hundred light-years.

A lunar scientific base also provides life scientists with a unique opportunity to study extensively biological processes in reduced gravity (one-sixth that of Earth) and in low magnetic fields. Genetic engineers can conduct their experiments in comfortable facilities that are nevertheless physically isolated from the Earth's biosphere. Exobiologists can experiment with new types of plants and microorganisms under a variety of simulated alien-world conditions. Genetically engineered "lunar plants" that were grown in special greenhouse facilities could become a major food source, while also supplementing the regeneration of a breathable atmosphere for the various lunar habitats.

The true impetus for large, permanent lunar settlements will most likely arise from the desire for economic gain—a time-honored stimulus that has driven much technical, social, and economic development on Earth. The ability to create useful products from native lunar materials will have a controlling influence on the overall rate of growth of the lunar civilization. Some early lunar products can now easily be identified. Lunar ice, especially when refined into pure water or dissociated into the important chemicals, hydrogen (H_2) and oxygen (O_2), represents the Moon's most important resource. Other important early lunar products include: (1) oxygen (extracted from lunar soils) for use as a propellant by orbital transfer vehicles traveling throughout cislunar space; (2) raw (i.e., bulk, minimally processed) lunar soil and rock materials for space radiation shielding; and (3) refined ceramic and metal products to support the construction of large structures and habitats in space.

The initial lunar base can be used to demonstrate industrial applications of native Moon resources and to operate small pilot factories that provide selected raw and finished products for use both on the Moon and in Earth orbit. Despite the actual distances involved, the cost of shipping a pound (kg) of "stuff" from the surface of the Moon to various locations in cislunar space may prove much cheaper than shipping the same "stuff" from the surface of Earth.

The Moon has large supplies of silicon, iron, aluminum, calcium, magnesium, titanium, and oxygen. Lunar soil and rock can be melted to make glass—in the form of fibers, slabs, tubes, and rods. Sintering (a process whereby a substance is formed into a coherent mass by heating but without melting) can produce lunar bricks and ceramic products. Iron metal can be melted and cast or converted to specially shaped forms using powder metallurgy. These lunar products would find a ready market as shielding materials, in habitat construction, in the development of large space facilities, and in electric-power generation and transmission systems.

Lunar mining operations and factories can be expanded to meet growing demands for lunar products throughout cislunar space. With the rise of lunar agriculture (accomplished in special enclosed facilities), the Moon may even become our "extraterrestrial breadbasket"—providing the majority of all food products consumed by humanity's extraterrestrial citizens.

One interesting space-commerce scenario involves an extensive lunar-surface mining operation that provides the required quantities of materials in a preprocessed condition to a giant space manufacturing complex located at Lagrangian libration point 4 or 5 (L4 or L5). These exported lunar materials would consist primarily of oxygen, silicon, aluminum, iron, magnesium, and calcium that were locked into a great variety of complex chemical compounds. It has often been suggested by space visionaries that the Moon will become the chief source of materials for space-based industries in the latter part of this century.

Numerous other tangible and intangible advantages of lunar settlements will accrue as a natural part of their creation and evolutionary development. For example, the high-technology discoveries that originate in a complex of unique lunar laboratories could be channeled directly into appropriate economic and technical sectors on Earth as "frontier" ideas, techniques, products, and so on. The permanent presence of people on another world (a world that looms large in the night sky) will suggest continuously an open world philosophy and a sense of cosmic destiny to the vast majority of humans who remain behind on the home planet. The human generation that decides to venture into cislunar space and to create permanent lunar settlements will long be admired not only for its great technical and intellectual achievements but also for its innovative cultural accomplishments. Finally, it is not too remote to speculate that the descendants of the first lunar settlers will become first the interplanetary, then the interstellar, portion of the human race. The Moon can be viewed as humanity's stepping stone to the universe.

✧ Mars Outpost and Surface Base Concepts

For automated Mars missions, the spacecraft and robotic surface rovers generally will be small and self-contained. For human expeditions to the surface of the Red Planet, however, two major requirements must be satisfied: life support (habitation) and surface transportation (mobility). Habitats, power supplies, and life-support systems will tend to be more complex in a permanent Martian surface base that must sustain human beings for years at a time. Surface mobility systems will also grow in complexity and sophistication as early Martian explorers and settlers travel tens to hundreds of miles (km) from their base camp. At a relatively early time in any Martian surface base program, the use of Martian resources to support the base must be tested vigorously and then quickly integrated in the development of an eventually self-sustaining surface infrastructure.

In one candidate scenario, the initial Martian habitats will resemble standardized lunar base (or space station) pressurized modules and would be transported from cislunar space to Mars in prefabricated condition by interplanetary nuclear-electric propulsion (NEP) cargo ships. These modules would then be configured and connected as needed on the surface of Mars and covered with about 3 feet (1 m) or so of Martian soil for protection against the lethal effects of solar-flare radiation or continuous exposure to cosmic rays on the planet's surface. Unlike Earth's atmosphere, the very thin Martian atmosphere does not shield very well against ionizing radiations from space.

Another midcentury Mars base concept involves an elaborate complex of habitation modules, power modules, central base work facilities, a greenhouse, a launch and landing complex, and even a robotic Mars airplane. The greenhouse on Mars would provide astronauts with some much-needed dietary variety. As an early Mars outpost grows into a

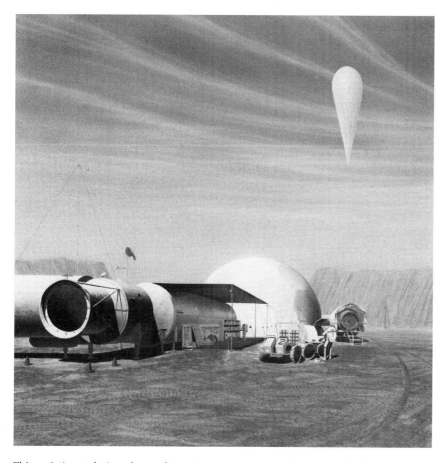

This artist's rendering shows the major components of one possible Mars outpost that could support up to seven astronauts while they explored the surface of the Red Planet. The main components are a habitat module, pressurized rover dock/equipment lock, airlocks, and a 52.5-foot- (16-m-) diameter, erectable (inflatable) habitat. Also appearing in the picture are a Mars balloon, an unpressurized rover, a storage work area, a geophysical experiment area, and a local area antenna. In the scenario depicted, many of the elements of this Mars outpost were derived from an earlier lunar test bed facility. *(NASA/JSC; artist, Mark Dowman of John Frassanito & Associates)*

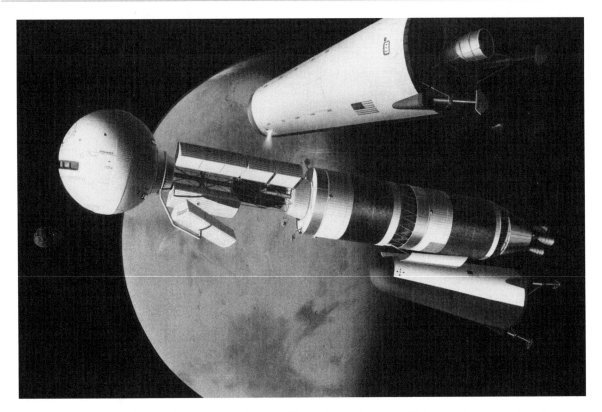

This artist's rendering shows a human-crewed nuclear thermal rocket–powered interplanetary cargo transfer vehicle that is on the way to the Jovian system and is being refueled in orbit around Mars near the Red Planet's moon Phobos. As the human race expands its presence out into the solar system, permanent settlements on Mars and refueling stations on its two moons could play a major "frontier town" role in the 22nd century. *(NASA: artist; Pat Rawlings, 1996)*

sufficiently large permanent human settlement, a system of greenhouses will be necessary to establish food self-sufficiency. In time, food grown at the Mars base could be used to supply human space-exploration missions that depart the Red Planet and travel into the asteroid belt and beyond.

✧ Planetary Engineering

Planetary engineering, or terraforming as it is sometimes called, is the large-scale modification or manipulation of the environment of a planet to make it more suitable for human habitation. In the case of Mars, human settlers would probably seek to make its atmosphere more dense and breathable by adding more oxygen. Early "Martians" would probably also attempt to alter the planet's harsh temperatures and to modify them to fit a more terrestrial thermal pattern. Venus represents an even larger challenge to the planetary engineer. Its current atmospheric pressure would have to be significantly reduced, its infernolike surface temperatures greatly diminished, the excessive amounts of carbon dioxide in its atmosphere reduced, and—perhaps the biggest task of all—its rotation rate increased to shorten the length of the solar day.

It should now be obvious that when scientists and engineers discuss planetary engineering projects, they are speaking of truly large, long-term projects. Typical time estimates for the total terraforming of a planet such as Mars or Venus range from centuries to a few millennia. However, we can also develop ecologically suitable enclaves or niches, especially on the Moon or Mars. Such localized planetary modification efforts could probably be accomplished within a few decades of project initiation.

Just what are the "tools" of planetary engineering? The planetary pioneers in the latter portions of this century will need at least the following, if they are to convert presently inhospitable worlds into new ecospheres that permit human habitation with little or no personal life-support equipment: first, and perhaps the most often overlooked, human ingenuity; second, a thorough knowledge of the physical processes of the particular planet or moon that is undergoing terraforming (especially the existence and location of environmental pressure points at which small modifications of the local energy or material balance can cause global environmental effects); third, the ability to manipulate large quantities

PLANETARY ENGINEERING

This is an artist's rendering of the domed-habitat approach to planetary engineering—an early space age vision of how to make the Moon and possibly Mars suitable for large-scale settlement by human pioneers decades after the Apollo Project. *(NASA)*

of energy; fourth, the ability to manipulate the surface or material composition of the planet; and fifth, the ability to move large quantities of extraterrestrial materials (for example, small asteroids, comets, or water-ice shipments from the Saturn rings) to any desired locations within heliocentric space.

One frequently suggested approach to planetary engineering is the use of biological techniques and agents to manipulate alien worlds into more desirable ecospheres. For example, scientists have proposed seeding the Venusian atmosphere with special microorganisms (such as genetically engineered algae) that are capable of converting excess carbon dioxide into free oxygen and combined carbon. This biological technique would not only provide a potentially more breathable Venusian atmosphere but it would also help to lower the currently intolerable surface temperatures by reducing the runaway greenhouse effect.

Other individuals have suggested the use of special vegetation (such as genetically engineered lichen, small plants, or scrubs) to help modify the polar regions on Mars. The use of specially engineered, survivable plants would reduce the albedo of these frigid regions by darkening the surface, thereby allowing more incident sunlight to be captured. In time, an increased amount of solar energy absorption would elevate global temperatures and cause melting of the long frozen volatiles, including water. This would raise the atmospheric pressure on Mars and possibly cause a greenhouse effect. With the polar caps melted, large quantities of liquid water would be available for transport to other regions of the planet. Perhaps one of the more interesting Martian projects late this century will be to construct a series of large irrigation canals.

Of course, there are other alternatives to help melt the Martian polar caps. The Martian settlers could decide to construct giant mirrors in orbit above the Red Planet. These mirrors would be used to concentrate and focus raw sunlight directly on the polar regions. Other scientists have suggested dismantling one of the Martian moons (Phobos or Deimos) or perhaps a small dark asteroid and then using its dust to darken the polar regions physically. This action would again lower the albedo and increase the absorption of incident sunlight.

Another approach to terraforming Mars is to use nonbiological replicating systems—that is, robots that can perform work as well as function as self-replicating systems. (The role and function of the self-replicating system in spreading life around in the solar system and beyond is discussed in a later section of this chapter.) These self-replicating machines probably will be able to survive more hostile environmental conditions than will genetically engineered microorganisms or plants.

To examine the scope and magnitude of this type of planetary engineering effort, we first assume that the Martian crust is mainly silicone dioxide (SiO_2) and then that a general purpose 100-ton, self-replicating system (SRS) "seed machine" can make a replica of itself on Mars in just one year. This SRS unit initially would make other units like itself, using native Martian raw materials. In the next phase of the planetary engineering project, these SRS units would be used to reduce SiO_2 into oxygen that is then released into the Martian atmosphere. In just 36 years from the arrival of the "seed machine," a silicon dioxide reduction capability would be available that could release up to 220,000 tons per second of pure oxygen into the thin atmosphere of the Red Planet. In only 60 years of operation, this array of SRS units would have produced and liberated 8.8×10^{17} pounds (4×10^{17} kg) of oxygen into the Martian environment. Assuming negligible leakage through the Martian exosphere, this is enough "free" oxygen to create a 0.1-bar (10-kPa) pressure, breathable atmosphere across the entire plant. This pressure level is roughly equivalent to the terrestrial atmosphere at an altitude of 9,840 feet (3,000 m).

What would be the environmental impact of all these mining operations on Mars? Scientists estimate that the total amount of material that must be excavated to terraform Mars is on the order of 2.2×10^{18} pounds (1×10^{18} kg) of silicon dioxide. This is enough soil to fill a surface depression 0.6 mile (1 km) deep and about 370 miles (600 km) in diameter. This is approximately the size of the crater Edom near the Martian equator. The future Martians might easily rationalize: Just one small hole for Mars, but a new ecosphere for all of us transplanted humans!

Asteroids can also play an interesting role in planetary engineering scenarios. People have suggested crashing one or two "small" asteroids into depressed areas on Mars (such as the Hellas Basin) to deepen and enlarge the depression instantly. The goal would be individual or multiple (connected) instant depressions about 6 miles (10 km) deep and 62 miles (l00 km) across. These human-caused impact craters would be deep enough to trap a denser atmosphere—allowing a small ecological enclave or niche to develop. Environmental conditions in such enclaves could range from typical polar conditions to perhaps something almost balmy.

Other would-be planetary engineers have suggested crashing asteroids into Venus to help increase its spin. If the asteroid hits Venus's surface at just the right angle and speed, it conceivably could help speed up the planet's rotation rate—greatly assisting any overall planetary engineering project. Unfortunately, if the asteroid is too small or too slow, it will have little or no effect; if it is too large or hits too fast, it could possibly shatter the planet.

It has also been proposed that several large-yield nuclear devices be used to disintegrate one or more small asteroids that had previously been maneuvered into orbits around Venus. This would create a giant dust and debris cloud that would encircle the planet and reduce the amount of incoming sunlight. This, in turn, would lower surface temperatures on the planet and allow the rocks to cool sufficiently to start to absorb carbon dioxide from the dense atmosphere of Venus.

Finally, others who are for large-scale planetary engineering projects have suggested mining the rings of Saturn for frozen volatiles, especially water-ice, and then transporting these large chunks of ice back into the inner solar system for use on Mars, the Moon, or Venus.

✧ Large Space Settlements—Hallmark of a Solar-System Civilization

The large space settlement is often viewed as the centerpiece of a grand technical vision, involving the construction of humanmade miniworlds that would result in the spread of life and civilization throughout the solar system. Long before the space age began, the British physicist and writer John Desmond Bernal (1910–71) speculated about the colonization of space and the construction of very large, spherical space settlements (now called Bernal spheres) in his futuristic 1929 work *The World, the Flesh and the Devil.* Although Bernal's use of the term *space colony* has yielded to the more politically acceptable expression *space settlement,* his basic idea of a large, self-sufficient human habitat in space has stimulated numerous space age era studies. These subsequent studies have spawned other interesting habitat concepts—some engineering extrapolations of Bernal's basic notion and others dramatically different in form or purpose.

Complementing with the long-range vision of constructing miniworlds in space is the notion of creatively harvesting the resources found there. Generally, when people think about outer space, visions of vast emptiness, devoid of anything useful, come to their minds. However, space is really a new frontier that is rich with resources, including an essentially unlimited supply of (solar) energy, a full range of raw materials, and an environment that is both special—such as high vacuum, orbital access to continuous microgravity, physical isolation from the terrestrial biosphere—and reasonably predictable, although large solar flares represent unpredictable, occasional threats.

Since the start of the space age, investigations of meteorites, the Moon, Mars, and several asteroids and comets have provided tantalizing hints about the rich mineral potential of the extraterrestrial environment. NASA's Apollo Project expeditions to the lunar surface established that the average lunar soil contains more than 90 percent of the material needed to construct a complicated space industrial facility. The soil in the lunar highlands is rich in anorthosite, a mineral suitable for the extraction of aluminum, silicon, and oxygen. Other lunar soils have been found to contain ore-bearing granules of ferrous metals, such as iron, nickel, titanium, and chromium. Iron can be concentrated from the lunar soil (called regolith) before the raw material is even refined simply by sweeping magnets over regolith to gather the iron granules that are scattered within.

Remote sensing data of the lunar surface obtained in the 1990s by the Department of Defense's *Clementine* spacecraft and NASA's *Lunar Prospector* spacecraft have encouraged some scientists to suggest that useful quantities of water-ice are trapped in Moon's perpetually shaded polar regions. If this postulation proves true, then "ice mines" on the Moon could provide both oxygen and hydrogen—vital resources for permanent lunar settlements and space industrial facilities. The Moon would be able both to export chemical propellants for propulsion systems and to make up (resupply) materials for the life-support systems of large human habitats that were constructed in cislunar space.

Its vast mineral-resource potential, frozen volatile reservoirs, and strategic location will make Mars a critical supply depot for human expansion into the mineral-rich asteroid belt and to the giant outer planets and their fascinating collection of resource-laden moons. Smart robot explorers will assist the first human settlers on Mars, enabling these Martians

This artist's rendering provides an exterior view of a large space settlement that is capable of supporting about 10,000 people in cislunar space. As envisioned in various NASA-sponsored studies that were performed in the late 1970s, the inhabitants of this type of space settlement would harvest materials from the Moon and possibly from near-Earth asteroids to construct large satellite power systems, which would then provide energy to Earth. *(NASA/Ames Research Center)*

pioneers to assess quickly and efficiently the full resource potential of their new world. As the early Martian bases mature into large permanent settlements, they will become economically self-sufficient by exporting propellants, life-support-system consumables, food, raw materials, and manufactured products to feed the next wave of human expansion to the outer regions of the solar system. Cargo spacecraft will routinely travel between cislunar space and Mars, carrying specialty items to eager consumer markets in both extraterrestrial locations.

The asteroids, especially Earth-crossing asteroids, represent another important category of space resources. Recent asteroid rendezvous missions and analysis of meteorites (many of which scientists believe originate from broken-up asteroids) indicate that carbonaceous (C-type) asteroids may contain up to 10 percent water, 6 percent carbon, significant amount of sulfur, and useful amounts of nitrogen. S-class asteroids, which are common near the inner edge of the main asteroid belt and among the Earth-crossing asteroids, may contain up to 30 percent free metals (alloys of iron, nickel, and cobalt, along with high concentrations of precious metals). E-class asteroids may be rich sources of titanium, magnesium, manganese, and other metals. Finally, chondrite asteroids, which are found among the Earth-crossing population, are believed to contain accessible amounts of nickel, perhaps more concentrated than the richest deposits found on Earth.

Using smart machines, especially self-replicating systems (discussed shortly), space settlers later in this century might be able to manipulate large quantities of extraterrestrial matter and move it to wherever it is needed in the solar system. Many of these space resources will be used as the feedstock for the orbiting and planetary-surface base industries that will form the basis of interplanetary trade and commerce. For example, atmospheric (aerostat) mining stations could be set up around Jupiter and Saturn, extracting such materials as hydrogen and helium—especially helium-3, an isotope of great potential value in nuclear fusion research and applications. Similarly, Venus could be mined for the carbon dioxide in its atmosphere, Europa for water, and Titan for hydrocarbons. Large fleets of robot spacecraft might even be used to gather chunks of water-ice from Saturn's ring system, while a sister fleet of robot vehicles extracts metals from the main asteroid belt. Even the nuclei of selected comets could be intercepted and mined for frozen volatiles, including water-ice. Finally, beyond the orbit of Neptune—the eighth and outermost major planet in the solar system—is the Kuiper belt and its population of thousands and thousands of icy planetesimals, which range in size from a few hundred feet (meters) in diameter to hundreds of miles (km) in diameter.

✧ Androcells Everywhere!

The androcell is a concept that the German-American rocket scientist Krafft A. Ehricke (1917–84) introduced in the 1970s. According to Ehricke, the androcell would be a very large, humanmade new world—an independent, self-contained human biosphere not located on any naturally existing celestial object. In his strategic vision, such humanmade miniworlds (or planetellas) would use mass far more efficiently than the natural worlds of the solar system, which formed out of the original solar nebular material some 4.6 billion years ago. The naturally formed terrestrial planets (Earth, Mars, Mercury, and Venus) and the wide variety of moons found throughout the solar system are essentially "solid" spherical objects of great mass. The surface gravity force on each of these solid-surface worlds results from the self-attraction of this very large quantity of matter. However, except for the first mile (kilometer) or so, the interior of each of the natural worlds is essentially "useless" from the perspective of human habitation.

Instead of large quantities of matter, the androcell would use rotation (centrifugal inertia) to provide variable levels of artificial gravity. The unusable solid interior of a natural celestial body is now replaced (through human ingenuity and engineering) with many useful inhabitable layers of airtight, habitable cylinders. In concept, inhabitants of an androcell would be able to enjoy a truly variable lifestyle in a multiple-gravity-level miniworld. There would be a maximum gravity level at the outer edges of the androcell, tapering off to essentially zero gravity in the inner cylinder levels closest to the central hub.

One especially important idea contained within Ehricke's overall visionary concept is that the androcell would not be tied to the Earth-Moon system. Rather—with its giant space-based factories, farms, and fleets of merchant spacecraft—the androcell would be free to seek political and economic development throughout heliocentric space. The inhabitants of a particular androcell might trade with Earth, the Moon, Mars, or with other androcells.

These giant space settlements, containing from 10,000 to as many as 100,000 or more people, represent the space age analogy to the city-states of ancient Greece. The multiple-gravity-level lifestyle would also encourage migration to and from settlements on other "natural" worlds—perhaps a terraformed Mars, an environmentally subdued Venus, or maybe even one of the larger moons of the giant outer planets. In essence, as implied by the name, the androcell represents the enabling technology for the cellular division of humanity. While the majority of the androcells might cluster in the continuously habitable zone (CHZ) of the solar system, those autonomous extraterrestrial city-states powered by nuclear energy would allow their inhabitants to pursue physically and culturally diverse lifestyles throughout the outer solar system.

Of course, the human race already has the initial, natural androcell—it is called "Spaceship Earth." In time, inhabitants of humankind's parent world will be able to use their technical skills and intelligence to fashion a series of such androcells or other very large habitable space structures throughout the solar system. As the number of these artificial space habitats grows, a swarm of settlements could eventually encircle the Sun, capturing and using all of its radiant energy output. At that point, humans' solar-system civilization will have created a Dyson sphere, making the next step of cosmic mitosis—migration to the stars—technically, economically, and socially feasible.

✧ Dyson Sphere

The Dyson sphere is a huge, artificial biosphere that is created around a star by an intelligent species as part of its technological growth and expansion within an alien solar system. This hypothesized giant structure would probably be formed by a swarm of artificial habitats and miniplanets that are capable of intercepting essentially all the radiant energy from the parent star. The captured radiant energy would be converted for use through a variety of techniques such as living plants, direct thermal-to-electric conversion devices, photovoltaic cells, and perhaps other exotic (as yet undiscovered) energy conversion techniques.

In response to the second law of thermodynamics, waste heat and unusable radiant energy would be rejected from the "cold" side of the Dyson sphere to outer space. From contemporary knowledge of engineering heat transfer, the heat rejection surfaces of the Dyson sphere might have a temperature that ranges from a lower limit of about -100°F (-73°C) to an upper limit of about 80°F (27°C). When observed at interstellar distances, this enormous collection of millions of artificial structures would offer a distinct thermal infrared signature. An unidentified celestial object that is one to two astronomical units in diameter and lies within the previously mentioned temperature range could be an alien civilization's Dyson sphere.

This hypothesized astroengineering project is an idea of the British-American theoretical physicist Freeman John Dyson (1923–). In essence, what Dyson has proposed is that advanced extraterrestrial societies, responding to Malthusian pressures, would eventually expand into their local solar system, ultimately harnessing the full extent of its energy and materials resources. Just how much growth does this type of expansion represent?

To explore this interesting concept further, scientists usually invoke the principle of mediocrity (that is, that conditions are pretty much the same throughout the universe) and then use humans' solar system as the model. The energy output from the Sun—a G2V-spectral class star—is approximately 4×10^{26} watts (joules per second [J/s]). For all practical purposes, scientists and engineers treat the Sun as a blackbody radiator at a temperature of approximately 9,980°F (5,800 K). The vast majority of the Sun's energy output occurs as electromagnetic radiation, predominantly in the wavelength range 0.3 to 0.7 micrometers (µm), which corresponds to visible light.

As an upper limit, the available mass in the solar system for very large-scale (astroengineering) construction projects may be taken as the mass of the planet Jupiter, some 4.4×10^{27} pounds (2×10^{27} kg). Contemporary energy consumption by human beings on Earth is about 10 terawatts (TW). The next step is to postulate just a 1 percent growth in terrestrial energy consumption per year. At this growth rate, within a mere three millennia, humans' energy consumption would reach the energy output of the Sun itself. Today, several billion human beings live in a single biosphere, planet Earth—with a total mass of some 11×10^{24} pounds (5×10^{24} kg). A few thousand years from now, the Sun could be surrounded by a swarm of habitats, containing trillions of human beings.

The Dyson sphere may be regarded as representing an upper limit for physical growth within the solar system. It is simply "the best human beings can do" from an energy and materials point of view in this particular corner of the galaxy. The vast majority of these hypothesized human-made habitats will most likely be located in the ecosphere, or continuously habitable zone (CHZ), around the Sun—that is, at a distance of about a one astronomical unit (AU) from our parent star. However, this does not preclude the possibility that other habitats, powered by nuclear fusion energy, might also be found scattered throughout the outer regions of a somewhat dismantled solar system. Such fusion-powered habitats could also represent the technical precursors to the first interstellar space arks—enormously large starships that carry significant numbers of human beings on one-way journeys to explore and then settle other star systems.

By using humankind's solar system and planetary civilization as a model, some exobiologists project that within a few millennia after the start of industrial development, an intelligent species could rise from the level of a planetary civilization (that is, a Kardashev Type I civilization) and eventually occupy a swarm of artificial habitats that completely surround their parent star, creating a fully matured Kardashev Type II civilization. Of course, these intelligent creatures might also elect to pursue interstellar travel and galactic migration, as opposed to completing the Dyson sphere within their home star system. This decision would represent the start of a Kardashev Type III civilization. (See chapter 8 for a discussion of extraterrestrial civilizations.)

It was further postulated by Dyson that the existence of such advanced extraterrestrial civilizations might be detected by the presence of a characteristic thermal infrared signature from very large objects in space that had dimensions of one to two astronomical units in diameter. Thus far, sophisticated, space-based infrared telescopes, like NASA's *Spitzer Space Telescope,* have not detected such unusual objects.

The Dyson sphere is certainly a grand, far-reaching concept. It is also quite appropriate for each reader to realize that the *International Space*

Station (ISS), currently being assembled in low Earth orbit, can be viewed as the very *first* habitat in the potential swarm of artificial structures that humans might eventually construct as part of their solar-system-level civilization.

✧ The Starship

Perhaps the most frequently used concept in science fiction to push, pull, and spread intelligent life around the galaxy is the starship—a space vehicle that is capable of traveling the great distances between star systems within a reasonable period of time. Building such a vehicle is a considerable engineering challenge because even the closest stars in the Milky Way galaxy are generally light-years apart. In this book, the word *starship* is used to describe interstellar spaceships that are capable of carrying intelligent beings to other star systems, while purely robotic interstellar space vehicles are called *interstellar probes.*

What are the performance requirements for a starship? First, and perhaps most important, the vessel should be capable of traveling at a significant fraction of the speed of light (*c*). Ten percent of the speed of light (0.1*c*) is often considered as the lowest acceptable speed for a starship, while cruising speeds of 0.9*c* and beyond are considered highly desirable. This optic velocity cruising capability is necessary to keep interstellar voyages to reasonable lengths of time, both for the home civilization and for the starship crew.

Consider, for example, a trip to the nearest star system, Alpha Centauri—a triple star system about 4.23 light-years away. At a cruising speed of 0.1*c*, it would take about 43 years just to get there and another 43 years to return. The time dilation effects of travel at these "relatively low" relativistic speeds would not help too much either since a ship's clock would register the passage of about 42.8 years versus a terrestrial ground elapse time of 43 years. In other words, the crew would age about 43 years during the journey to Alpha Centauri. If the expedition started with 20-year-old crewmembers, who departing from the outer regions of the solar system in the year 2100 traveled at a constant cruising speed of 0.1*c*, the crew would be approximately 63 years old when they reached the Alpha Centauri star system—some 43 years later in 2143. The return journey would be even more dismal. Any surviving crew members would be 106 years old when the ship returned to the outer fringes of the solar system in the year 2186. Most, if not all, of the crew would probably have died of old age or of boredom.

A starship should also provide a comfortable living environment for the crew and the passengers (in the case of an interstellar ark). Living in a relatively small, isolated, and confined habitat for a few decades to perhaps a few centuries can certainly overstress even the most psychologically adaptable individuals and their progeny. One common technique used in science fiction to avoid this crew stress problem is to have all or most of the human crew placed in some form of "suspended animation"—while the vehicle travels through the interstellar void, tended by a ship's company of smart robots.

Any properly designed starship must also provide an adequate amount of radiation protection for the crew, passengers, biological materials, and sensitive electronic equipment. Interstellar space is permeated with galactic cosmic rays. Nuclear radiation leakage from an advanced thermonuclear fusion engine or a matter/antimatter engine (photon rocket) must also be prevented from entering the habitation compartment. In addition, the crew will have to be protected from nuclear radiation showers that are produced when a starship's hull, traveling at near light speed, slams into interstellar molecules, dust, or gas. For example, a single proton (assumed for mathematical convenience as being "stationary" in this simple example) that encounters a starship moving at 90 percent of the speed of light ($0.9c$) would appear to those on board like a one gigaelectron volt (GeV) proton being accelerated at them. Imagine traveling for years at the beam output end of a very high-energy particle accelerator. Without proper deflectors or shielding, survival in the crew compartment after such an enormous dose of nuclear radiation is doubtful.

To function truly as a starship, the vessel must be able to cruise at will, light-years from its home star system. The starship must also be able to accelerate to significant fractions of the speed of light, cruise at these near optic velocities, and then decelerate to explore a new star system or to investigate a derelict alien spaceship that has been found adrift in the depths of interstellar space.

This chapter does not discuss the enormous difficulties of navigating through interstellar space at near light velocities. It will be sufficient to mention here simply that when the crew "looks" forward at near light speeds, things appear blueshifted; while when they look aft (backward), things appear redshifted. The starship and its crew must be able to find their way independently from one location in the Milky Way galaxy to another.

What appears to be the major technology needed to make the starship a credible part of humans' extraterrestrial civilization is an effective propulsion system. Interstellar class propulsion technology is the key to the galaxy for any emerging civilization that has mastered spaceflight within and to the limits of its own star system. Despite the tremendous engineering difficulties that are associated with the development of a starship propulsion system, several concepts have been proposed. These include the pulsed nuclear-fission engine (Project Orion concept), the pulsed

An artist's rendering of an interstellar ramjet. The conceptual starship is characterized by a huge (39-square-mile [100-km²]) scoop to capture interstellar hydrogen for use as fuel in sustaining the thermonuclear reactions that propel the vehicle to upward of 90 percent of light speed. *(NASA/Marshall Space Flight Center)*

nuclear-fusion concept (Project Daedalus study), the interstellar nuclear ramjet, and the photon rocket.

Unfortunately, in terms of the way scientists and engineers currently understand of the laws of physics, all known phenomena and mechanisms that might be used to power a starship are either not energetic enough or simply are entirely beyond today's level of engineering technology. In fact, starship propulsion systems appear to require a level of space technology that is not envisioned for several decades, if not a century or more. Perhaps there will be major new insights as to how scientists perceive the physical laws of the universe, or perhaps engineers will develop better ways to manipulate energy and matter. But until such breakthroughs actually occur (if they ever do), the notion of human beings traveling in a starship to another star system must remain in the realm of future dreams and science fiction.

✦ Designing an Interstellar Probe

If the development of human crew-carrying starships proves to be an impossible technical feat during the next few centuries, the human race can still search other star systems for alien life and spread terrestrial, carbon-based life around the galaxy with the help of sophisticated robot spacecraft. A modest effort might even be started at the end of this century with a decision by humans to launch one or more interstellar probes. Then, sometime in the 22nd century, our descendants might make innovative use of self-replicating system (SRS) technology in robot spacecraft based "life-detecting" and "life-spreading" exploration campaigns that extend eventually across the entire galaxy.

This artist's rendering shows the human race's first interstellar robot probe departing the solar system (ca. 2075) on an epic journey of scientific exploration. *(NASA)*

An interstellar probe is a highly automated robot spacecraft sent from one solar system to explore another star system. Most likely this type of probe would make use of very smart machine systems that are capable of operating autonomously for decades or centuries.

Once the robot probe arrives at a new star system, it would begin a detailed exploration procedure. The target star system is scanned for possible life-bearing planets, and if any are detected, they become the object of more intense scientific investigations. Data collected by the probe (which also serves as a mother spacecraft) and any miniprobes (deployed to explore individual objects of interest within the new star system) are transmitted back to Earth. After years of travel, these signals eventually are intercepted and analyzed by terrestrial scientists. Over time, such probes provide a continuous stream of interesting, unpredictable discoveries that enrich knowledge of the nearby star systems and any life forms that might share the galaxy with the human race.

The robot interstellar probe could also be designed to carry a payload of specially engineered microorganisms, spores, and bacteria. If the robot probe encounters ecologically suitable planets on which life has not yet evolved, then it could make the decision to "seed" such barren but potentially fertile worlds with primitive life-forms. In that way, human beings (in partnership with their robot probes) would not only be exploring neighboring star systems but would be participating in the spreading of life itself through some portion of the Milky Way galaxy.

Long-range strategic planners in the aerospace field have examined some of the engineering and operational requirements of the first interstellar probe, as might be launched at the end of this century to a nearby star—most likely one within 10 light-years distance or less. Some of these challenging requirements (all of which exceed current levels of technology by one or two orders of magnitude) are briefly mentioned here. The interstellar probe must be capable of sustained, autonomous operation for more than 100 years. The robot spacecraft must be capable of managing its own health—that is, being able to anticipate or predict a potential problem, detect an emerging abnormality, and then preventing or correcting the situation. For example, if a subsystem were about to overheat (but has not yet exceeded thermal design limits), the smart robot probe would redirect operations and adjust the thermal control system to avoid the potentially serious overheating condition.

The first interstellar robot probe must have a very high level of machine intelligence and be capable of exercising fault management through repair, redundancy, and workarounds without any human guidance or assistance. The smart robot must also be able to manage its onboard resources carefully, supervising the generation and distribution of electric power, allocating the use of consumables, deciding when and where to commit emergency reserves and a limited supply of spare parts and components. The main onboard computer (or machine brain) of the probe must exercise data management skills and be capable of an inductive response to unknown or unanticipated environmental changes. When faced with unknown difficulties or opportunities, the robot probe must be able to modify the mission plan and to generate new tasks.

During one such hypothetical mission, long-range sensors on board the probe might discover that the hot-Jupiter-type extrasolar planet within the target star system has a large (previously unknown) moon with an atmosphere and a liquid-water ocean. Instead of sending its last miniprobe ahead to investigate the extrasolar planet, the smart robot mother spacecraft makes a decision to release its last miniprobe to perform close-up measurements of this interesting moon. Since the mother spacecraft is more than eight light-years from Earth when the (hypothesized) discovery is made, the decision to change the mission plan must be made exclusively by the robot spacecraft, which is less than a few light-days away from the encounter. Sending a message back to Earth and asking for instructions would take more than 16 years (for round-trip communications), and by then the interstellar probe would have completely passed through the target star system and disappeared into the interstellar void.

Similarly, instruments on board the interstellar probe (regarded here as the mother spacecraft) and its supporting cadre of miniprobes must be capable of deductive and inductive learning so as to adjust how measurements are taken in response to unfolding opportunities, feedback, and unanticipated values (high and low). Some of the greatest scientific discoveries on Earth happened because of an accidental measurement or unanticipated reading.

So, the instruments on board the robot probe must be capable of exercising some "artificial thinking" level of curious inquiry and then be able to respond to unanticipated but quite significant new findings. The robot probe should also have a level of machine intelligence that is capable of knowing and appreciating when newly acquired data is very significant. This is a difficult task for human scientists, who often overlook the most significant pieces of data in an experiment or observation. To ask a robot's mechanical brain to respond "eureka" (I've found it) at a moment of a great discovery is pushing machine intelligence well beyond the technical horizon that is projected for the next few decades. Yet, if the human race is going to make significant discoveries using robot interstellar probes, that is precisely what these advanced exploring machines must be capable of doing.

From a purely engineering perspective, the interstellar robot probe should consist of low-density, high-strength materials to minimize propulsion requirements. Remember that to keep a mission to the nearby stars within 100 years or so duration, the robot spacecraft should be capable of cruising at about one-tenth the speed of light (or more). Any less than that and a star-probe mission to even the nearest stars would take several centuries. The great-great-great-grandchildren of the probe engineers would somehow have to remain interested in receiving the signals from the (probably long-forgotten) probe. So, these early interstellar probe missions (using advanced, but nonreplicating technology) will most likely take a 100 years or less.

The materials used on the outside of the robot probe must maintain their integrity for more than a century, even when subjected to harsh, deep space conditions—such as ionizing radiation, cold, vacuum, and interstellar dust. The structure of the robot spacecraft should be capable of autonomous reconfiguration. The power system must be able to provide reliable base power (typically at a level of 100 kilowatts-electric up to possibly one megawatt-electric) on an autonomous and self-maintaining basis for more than 100 years. Finally, the star probe must be capable of autonomous data collection, assessment, storage, and communications (back to Earth) from a wide variety of scientific instruments and onboard spacecraft state-of-health sensors.

Some of the intriguing challenges in information technology include the proper calibration of instruments and the collection of data during a period of years after decades of sensor dormancy. The robot probe must be able to transmit data back to Earth for distances ranging from 4.5 to 10.0 light-years, or more. Finally, after decades of handling modest levels of data, the spacecraft's information systems must be capable of handling a gigantic burst of incoming data as the robot probe and its miniprobes encounter the target star system.

While this section and the remaining sections of the chapter discuss robot spacecraft that are launched by the human race to explore nearby star systems, the reader should not lose sight of the fact that the reverse circumstance is also a distinct possibility. An alien civilization, perhaps 10 to 15 light-years distant from Earth and about a century ahead of the human race in technology, could right now be in the process of sending its own robot probes to investigate humans' solar system and home planet.

✧ The Theory and Operation of Self-Replicating Systems

The brilliant Hungarian-American mathematician John von Neumann (1903–57) was the first person to consider the problem of self-replicating systems seriously. His book on the subject, *Theory of Self-reproducing Automata,* was edited by a colleague, Arthur W. Burks (1915–   ), and was published posthumously in 1966—almost a decade after von Neumann's untimely death due to cancer.

Von Neumann became interested in the study of automatic replication as part of his wide-ranging interests in complicated machines. His work during the World War II Manhattan Project (the top secret U.S. atomic bomb project) led him into automatic computing. Through this association, he became fascinated with the idea of large complex computing machines. In fact, he invented the scheme that is used today in the great majority of general-purpose digital computers—the von Neumann concept of serial processing stored-program—and that is also referred to as the von Neumann machine.

In 1945, von Neumann drafted a report in which he introduced the concept of the stored-program computer. He also recognized that base 2 represented a considerable gain in computer design simplicity over the base-10 approach, which had been used in the world's first working electronic calculator and digital computer, the Electronic Numerical Integrator and Computer (ENIAC). It was completed in 1946 and contained 18,000 vacuum tubes. While it was a major step forward in the evolution of "thinking machines," ENIAC stored and manipulated numbers in base 10. Von Neumann's suggestion of using the base 2 allowed circuits in the digital computer to assume only two states: on or off or 0 or 1 (in binary notation).

Following his pioneering work in computer science, of which he is one of the founding fathers, von Neumann decided to tackle the larger problem of developing a self-replicating machine. The theory of automata provided him a convenient synthesis of his early efforts in logic and proof theory and his more recent efforts (during and after World War II) on large-scale electronic computers. Von Neumann continued to work on the intriguing idea of a self-replicating machine and its implications until his death in 1957.

Von Neumann actually conceived of several types of self-replicating systems, which he called the kinematic machine, the cellular machine, the neuron-type machine, the continuous machine, and the probabilistic machine. Unfortunately, he was only able to develop a very informal description of the kinematic machine before his death in 1957.

The kinematic machine is the most frequently discussed von Neumann–type self-replicating system. For this type of SRS, von Neumann envisioned a machine residing in a "sea of spare parts." The kinematic machine would have a memory tape that instructed the device to go through certain mechanical procedures. Using manipulator arms and its ability to move around, this type of SRS would gather and assemble parts. The stored computer program would instruct the machine to reach out and pick up a certain part and then go through an identification and evaluation routine to determine whether the part selected was or was not called for by the master tape. (Note: in von Neumann's day, microprocessors, minicomputers, floppy disks, CD-ROMs, and multigigabyte-capacity laptop computers did not exist.) If the component picked up by the manipulator arm did not meet the selection criteria, it was tossed back into the parts bin (that is, back into the "sea of parts.") The process would continue until the required part was found, and then an assembly operation would be performed. In this way, von Neumann's kinematic SRS eventually would make a complete replica of itself—without, however, understanding what it was doing. When the duplicate was physically completed, the parent machine would make a copy of its own memory tape on the (initially) blank tape of its offspring. The last instruction on the parent's machine tape would be to activate the tape of its mechanical progeny. The offspring kinematic SRS could then start searching the "sea of parts" for components to build yet another generation of SRS units.

In dealing with his self-replicating system concepts, von Neumann concluded that these machines should include the following characteristics and capabilities: (1) logical universality; (2) construction capability; (3) constructional universality; and (4) self-replication. Logical universality is simply the device's ability to function as a general-purpose computer. To be able to make copies of itself, a machine must be capable of manipulating information, energy, and materials. This is what is meant by the term *construction capability*. The closely related term *constructional universality* is a characteristic that implies the machine's ability to manufacture any of the finite-sized machines that can be built from a finite number of different parts, which are available from an indefinitely large supply. The characteristic of self-reproduction means that the original machine, given a sufficient number of component parts (of which it is made) and sufficient instructions, can make additional replicas of itself.

One characteristic of SRS devices that von Neumann did not address but that has been addressed by subsequent investigators is the concept of evolution. In a long sequence of machines that make machines like themselves, can successive robot generations learn how to make themselves better machines? Robot engineers and artificial intelligence experts are exploring this intriguing issue as part of the larger question of thinking machines that are self-aware.

Can robots be made smart and alert enough to learn from the experiences that are encountered in daily operations and thus improve their performance? If so, will such improvements simply reflect a primitive level of machine learning, or will the smart machines somehow begin to develop an internal sense of "knowing" that they know? If and when this ever occurs, the smart robot will begin to mimic the consciousness of its human creators. Some AI researchers like to speculate boldly that an advanced "thinking" robot in the distant future could be capable

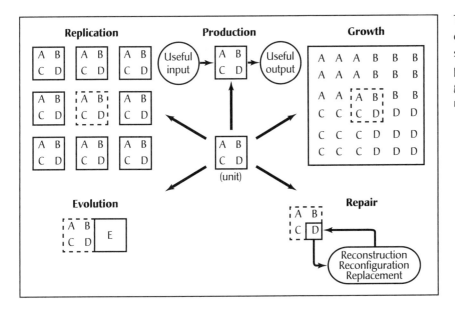

The five general classes of self-replicating system (SRS) behavior: production, replication, growth, evolution, and repair. *(NASA)*

of formulating the famous philosophical postulate of René Descartes (1596–1650): *"Cogito, ergo sum"* (I think, therefore I am). An SRS unit that exhibits the behavior of evolution might certainly be capable of achieving some form of machine self-awareness.

From von Neumann's work and the more recent work of other investigators, five broad classes of SRS behavior have been suggested:

1. *Production.* The generation of useful output from useful input. In the production process, the unit machine remains unchanged. Production is a simple behavior demonstrated by all working machines, including SRS devices.
2. *Replication.* The complete manufacture of a physical copy of the original machine unit by the machine unit itself.
3. *Growth.* An increase in the mass of the original machine unit by its own actions, while still retaining the integrity of its original design. For example, the machine might add an additional set of storage compartments in which to keep a larger supply of parts or constituent materials.
4. *Evolution.* An increase in the complexity of the unit machine's function or structure. This is accomplished by additions or deletions to existing subsystems or by changing the characteristics of these subsystems.
5. *Repair.* Any operation performed by a unit machine on itself that helps reconstruct, reconfigure, or replace existing subsystems—but does not change the SRS unit population, the original unit mass, or its functional complexity.

In theory, replicating systems can be designed to exhibit any or all of these machine behaviors. When such machines are actually built, however, a particular SRS unit will most likely emphasize just one or several kinds of machine behavior, even if it were capable of exhibiting all of them. For example, the fully autonomous, general-purpose self-replicating lunar factory, proposed in 1980 by Georg von Tiesenhausen and Wesley A. Darbo of the Marshall Space Flight Center (MSFC), is an SRS design concept that is intended for unit replication. There are four major subsystems that make up this proposed SRS unit. First, a materials processing subsystem gathers raw materials from its extraterrestrial environment (the lunar surface) and prepares industrial feedstock. Next, a parts production subsystem uses this feedstock to manufacture other parts or entire machines.

At this point, the conceptual SRS unit has two basic outputs. Parts may flow to the universal constructor (UC) subsystem, where they are used to make a new SRS unit (this is replication), or parts may flow to a production facility subsystem, where they are made into commercially useful products. This self-replicating lunar factory has other secondary subsystems, such as a materials depot, parts depot, power supply, and command and control center.

The universal constructor (UC) manufactures complete SRS units that are exact replicas of the original SRS unit. Each replica can then

An artist's rendering illustrating the general structure and basic components of a conceptual self-replicating lunar factory. (NASA)

make additional replicas of itself until a preselected SRS unit population is achieved. The universal constructor would retain overall command and control (C&C) responsibilities for its own SRS unit as well as for its mechanical progeny—until, at least, the C&C functions themselves have been duplicated and transferred to the new units. To avoid cases of uncontrollable exponential growth of such SRS units in some planetary resource environments, the human masters of these devices may reserve the final step of the C&C transfer function to themselves or so design the SRS units such that the final C&C transfer function from machine to machine can be overridden by external human commands.

✧ Extraterrestrial Impact of Self-Replicating Systems

The issue of closure (total self-sufficiency) is one of the fundamental problems in designing self-replicating systems. In an arbitrary SRS unit, there are three basic requirements necessary to achieve closure: (1) matter closure; (2) energy closure; and (3) information closure. In the case of matter closure, engineers ask: Can the SRS unit manipulate matter in all the ways that are needed for complete self-construction? If not, the SRS unit has not achieved matter or material closure. Similarly, engineers ask whether the SRS unit can generate a sufficient amount of energy that is needed and in the proper form to power the processes that are needed for self-construction. Again, if the answer is no, then the SRS unit has not achieved energy closure. Finally, engineers must ask: Does the SRS unit successfully command and control all the processes that are necessary for complete self-construction? If not, information closure has not been achieved.

If the machine device is only partially self-replicating, then engineers say that only partial closure of the system has occurred. In this case, some essential energy, or information must be provided from external sources, or else the machine system would fail to reproduce itself.

Just what are the applications of self-replicating systems? The early development of SRS technology for use on Earth and in space should trigger an era of superautomation that will transform most terrestrial industries and lay the foundation for efficient space-based industries. One interesting machine is called the Santa Claus machine—originally suggested and named by the American physicist Theodore Taylor (1925–2004). In this particular version of an SRS unit, a fully automatic mining, refining, and manufacturing facility gathers scoopfuls of terrestrial or extraterrestrial materials. It then processes these raw materials by means of a giant mass spectrograph that has huge superconducting magnets. The material is converted into an ionized atomic beam and sorted into stockpiles of basic elements, atom by atom. To manufacture any item, the Santa Claus machine selects the necessary materials from its stockpile, vaporizes them, and injects them into a mold that changes the materials into the desired item. Instructions for manufacturing, including directions on adapting new processes and replication, are stored in a giant computer within the Santa Claus machine. If the product demands becomes excessive, the Santa Claus machine would simply reproduce itself.

SRS units might be used in very large space-construction projects (such as lunar mining operations) to facilitate and accelerate the exploitation of extraterrestrial resources and to make possible feats of planetary engineering. For example, mission planners could deploy a seed SRS unit on Mars as a prelude to permanent human habitation. This machine would use local Martian resources to manufacture a large number of robot explorer vehicles automatically. This armada of vehicles would be disbursed over the surface of the Red Planet, searching for the minerals and frozen volatiles that are needed in the establishment of a Martian civilization. In just a few years, a population of some 1,000 to 10,000 smart machines could scurry across the planet, completely exploring its entire surface and preparing the way for permanent human settlements.

Replicating systems would also make possible large-scale interplanetary mining operations. Extraterrestrial materials could be discovered, mapped, and mined, using teams of surface and subsurface prospector robots that were manufactured in large quantities in an SRS factory complex. Raw materials would be mined by hundreds of machines and then sent wherever they were needed in heliocentric space. Some of the raw materials might even be refined in transit, with the waste slag being used as the reaction mass for an advanced propulsion system.

Atmospheric mining stations could be set up at many interesting and profitable locations throughout the solar system. For example, Jupiter and Saturn could have their atmospheres mined for hydrogen, helium (including the very valuable isotope, helium-3), and hydrocarbons, using aerostats. Cloud-enshrouded Venus might be mined for carbon dioxide, Europa for water, and Titan for hydrocarbons. Intercepting and mining comets with fleets of robot spacecraft might also yield large quantities of useful volatiles. Similar mechanized space armadas could mine water-ice from Saturn's ring system. All of these smart space robot devices would be mass-produced by seed SRS units. Extensive mining operations in the main asteroid belt would yield large quantities of heavy metals. Using extraterrestrial materials, these replicating machines could, in principle, manufacture huge mining or processing plants or even ground-to-orbit or interplanetary vehicles. This large-scale manipulation of the solar system's material resources would occur in a very short period of time, perhaps within one or two decades of the initial introduction of replicating-machine technology.

From the viewpoint of a solar-system civilization, perhaps the most exciting consequence of the self-replicating system is that it would provide a technological pathway for organizing potentially infinite quantities of matter. Large reservoirs of extraterrestrial matter might be gathered and organized to create an ever-widening human presence throughout heliocentric space. Self-replicating space stations, space settlements, and domed cities on certain alien worlds of the solar system would provide a diversity of environmental niches never before experienced in the history of the human race.

The SRS unit would provide such a large amplification of matter-manipulating capability that it would be possible for humans to start to consider seriously planetary engineering (or terraforming) strategies for the Moon, Mars, Venus, and certain other alien worlds. In time, advanced self-replicating systems could be used in the 22nd century as part of human's solar-system civilization to perform incredible feats of astroengineering. The harnessing of the total radiant energy output of the Sun, through the robot-assisted construction of a Dyson sphere, is an exciting example of one large-scale astroengineering project that might be undertaken.

Advanced SRS technology also appears to be the key to human exploration and expansion beyond the very confines of the solar system. This application illustrates the fantastic power and virtually limitless potential of the SRS concept.

It seems logical that before humans travel into the interstellar void, smart robot probes will be sent ahead as scouts. Interstellar distances are so large and search volumes so vast that self-replicating probes (sometimes referred to as von Neumann probes) represent a highly desirable, if not totally essential, approach to performing detailed studies of a large number of other star systems, including the search for extraterrestrial life.

One speculative study on galactic exploration suggests that search patterns beyond the 100 nearest stars probably would be optimized by the use of SRS probes. In fact, reproductive probes might permit the direct reconnaissance of the nearest one million stars in about 10,000 years and the entire Milky Way Galaxy in less than one million years—starting with a total investment by the human race of just one self-replicating interstellar robot spacecraft.

Of course, the problems of tracking, controlling, and assimilating all the data sent back to the home star system by an exponentially growing number of robot probes is simply staggering. Humans might avoid some of these problems by sending only very smart machines that are capable of greatly distilling the information gathered and transmitting only the most significant data, suitably abstracted, back to Earth. Robot engineers might also devise some type of command and control hierarchy in which each robot probe only communicates with its parent. Thus, a chain of ancestral repeater stations could be used to control the flow of messages and exploration reports through interstellar space as this bubble of machines pushes out into the galaxy.

Imagine the exciting chain reaction that might occur as one or two of the leading probes encountered an intelligent alien race. If the alien race proved hostile, an interstellar alarm would be issued, taking years to ripple back across the interstellar void at the speed of light, repeater station by repeater station, until Earth received notification. Would future citizens of Earth respond by sending more sophisticated, possibly predator, robot probes to that area of the galaxy? Perhaps, our descendants would decide instead simply to quarantine the belligerent species by positioning warning beacons all around the region. These warning beacons would signal any approaching self-replicating robot probes to swing clear of the hostile alien encounter zone. Robotic defensive systems might also be sent to enforce the galactic quarantine, keeping the hostile species confined to a small volume of the galaxy.

In time, as first hypothesized early in the 20th century by the U.S. rocket expert Robert Hutchings Goddard (1882–1945), giant space arks, representing an advanced level of synthesis between human crew and robot crew, could depart from the solar system and journey through the interstellar void. On reaching another star system that contained suitable material resources, the space ark itself would undergo replication. The human passengers (perhaps several generations of humans beyond the initial crew that departed the solar system) could then redistribute themselves between the parent space ark, offspring space arks, and any suitable extrasolar planets found orbiting that particular star. In a sense, the original space ark would serve as a self-replicating "Noah's Ark" for the human race and any terrestrial life-forms carried on board the giant, mobile habitat. This dispersal of conscious intelligence (that is, intelligent human life) to a variety of ecological niches within other star systems would ensure that not even disaster on a cosmic scale, such as the death of the Sun, could threaten the complete destruction of the human species and all human accomplishments. Astronomers predict that the death of the Sun will take place in about five billion years—when our parent star runs out of hydrogen for fusion in its core, leaves the main sequence, expands into a red giant, and ultimately collapses into a white dwarf.

The self-replicating space ark would enable human beings literally to send a wave of consciousness and carbon-based life (as we know it) into the galaxy. Sometimes referred to as the greening of the galaxy, this propagating wave of human intelligence in partnership with advanced machine intelligence would promote a golden age of interstellar development—at least within a portion of the Milky Way galaxy. How far this wave of conscious intelligence would propagate out into the galaxy is anyone's guess at this point.

✧ Control of Self-Replicating Systems

Whenever engineers discuss the technology and role of self-replicating systems, their conversations inevitably turn to the interesting question: What happens if a self-replicating system (SRS) becomes out of control? Before human beings seed the solar system or interstellar space with even a single SRS unit, engineers and mission planners should know how to pull an SRS unit's plug if things grow out of control. Some engineers and scientists have already raised a very legitimate concern about SRS technology. Another question that robot engineers often encounter concerning SRS technology is whether smart machines represent a long-range threat to human life. In particular, will machines evolve with such advanced levels of artificial intelligence that they become the main resource competitors and adversaries of human beings—whether the ultrasmart machines can replicate or not? Even in the absence of advanced levels of machine intelligence that mimic human intelligence, the self-replicating system might represent a threat just through its potential for uncontrollable exponential growth.

These questions can no longer remain entirely in the realm of science fiction. Engineers and scientists must start examining the technical and social implications of developing advanced machine intelligences and self-replicating systems *before* they bring such systems into existence. Failure to engage in such prudent and reasonable forethoughts could promote a future situation (now very popular in science fiction) in which human beings find themselves in a mortal conflict over planetary (or solar system) resources with their own intelligent machine creations.

Of course, human beings definitely need smart machines to improve life on Earth, to explore the solar system, to create a solar-system civilization, and to probe the neighboring stars. So engineers and scientists should proceed with the development of smart machines, but they should also temper these efforts with safeguards to avoid the ultimate undesirable future situation in which the machines turn against their human masters and eventually enslave or exterminate them. In 1942, the science-fact-fiction writer Isaac Asimov (1920–92) suggested a set of rules for robot behavior in his science-fiction story "Runaround," which appeared in *Astounding* magazine.

Over the years, Asimov's laws have become part of the cult and culture of modern robotics. They are: (Asimov's First Law of Robotics) "A robot may not injure a human being, or, through inaction, allow a human being to come to harm;" (Asimov's Second Law of Robotics) "A robot must obey the orders given it by human beings except where such orders would conflict with the first law;" and (Asimov's Third Law) "A robot must protect its own existence as long as such protection does not conflict with the first or second law." The message within these so-called laws represents a good starting point in developing benevolent, people-safe, smart machines.

However, any machine that is sophisticated enough to survive and reproduce in largely unstructured environments would probably also be capable of performing a certain degree of self-reprogramming, or automatic improvement (that is, have the machine behavior of evolution). An intelligent SRS unit eventually might be able to program itself around any rules of behavior that were stored in its memory by its human creators. As it learned more about its environment, the smart SRS unit might decide to modify its behavior patterns to better suit its own needs. If this very smart SRS unit really "enjoys" being a machine and making (and perhaps improving) other machines, then when faced with a situation in which it must save a human master's life at the cost of its own, the smart machine may decide to simply shut down instead of performing the life-saving task that it was preprogrammed to do. Thus, while it did not harm the endangered human being, it did not help the person out of danger either. Viewed on a larger scale, an entire population of "people-safe" robots passively might allow the human race to collapse and then fill in the intelligence void in this corner of the galaxy.

Science fiction contains many interesting stories about robots, androids, and even computers turning on their human builders. The conflict between the human astronaut crew and the interplanetary spaceship's feisty computer, HAL, in Arthur C. Clarke and Stanley Kubrick's cinematic masterpiece *2001: A Space Odyssey* is an incomparable example. The purpose of this brief discussion is not to invoke a Luddite-type response against the development of very smart robots but only to suggest that such exciting research and engineering activities be tempered by some forethought concerning the potential technical and social impact of these developments both here on Earth and throughout the galaxy.

One or all of the following techniques might control an SRS population in space. First, the human builders could implant machine-genetic instructions (deeply embedded computer code) that contained a hidden or secret cutoff command. This cutoff command would be activated automatically after the SRS units had undergone a predetermined number of replications. For example, after each machine replica is made, one regeneration command could be deleted—until, at last, the entire replication process is terminated with the construction of the last (predetermined) replica.

Second, a special signal from Earth at some predetermined emergency frequency might be used to shut down individual, selected groups, or all SRS units at any time. This approach is like having an emergency stop button which when pressed by a human being causes the affected SRS units to cease all activities and go immediately into a safe, hibernation posture. Many modern machines have either an emergency stop button, flow cut-off valve, heat limit switch, or master circuit breaker. The signal activated "all-stop" button on an SRS unit would just be a more sophisticated version of this engineered safety device.

For low-mass SRS units (perhaps in the 200-pound [100-kg] to 10,000-pound [4,500-kg] class), population control might prove more difficult because of the shorter replication times when compared with much-larger-mass SRS factory units. To keep these mechanical critters in line, human managers might decide to use a predator robot. The predator robot would be programmed to attack and destroy only the type of SRS units whose populations were out of control due to some malfunction or other. Robot engineers have also considered SRS unit population control through the use of a universal destructor (UD). This machine would be capable of taking apart any other machine that it encountered. The universal destructor would recover any information found in the prey robot's memory prior to recycling the prey machine's parts. Wildlife managers on Earth today use (biological) predator species to keep animal populations in balance; similarly, space robot managers in the future could use a linear supply of nonreplicating machine predators to control an exponentially growing population of misbehaving SRS units.

Engineers might also design the initial SRS units to be sensitive to population density. Whenever the smart robots sensed overcrowding or overpopulation, the machines could lose their ability to replicate (that is, become mechanically infertile), stop their operations, and go into a hibernation state—or perhaps (like lemmings on Earth) report to a central facility for disassembly. Unfortunately, SRS units might mimic the behavior patterns of their human creators too closely, so without preprogrammed behavior safeguards, overcrowding could force such intelligent machines to compete among themselves for dwindling supplies of resources (terrestrial or extraterrestrial). Dueling, mechanical cannibalism, or even some highly organized form of robot-versus-robot conflict might result.

Hopefully, future human engineers and scientists will create smart machines that mimic only the best characteristics of the human species, for it is only in partnership with very smart and well-behaved self-replicating systems that human beings can some day hope to send a wave of life, conscious intelligence, and organization through the Milky Way galaxy.

In the very long term, there appear to be two general pathways for the human species: either human beings are a very important biological stage in the overall evolutionary scheme of matter and energy in the universe, or else humans are an evolutionary dead end. If the human race decides to limit itself to just one planet (Earth), a natural disaster or our own foolhardiness will almost certainly terminate the species—perhaps in just a few centuries or a few millennia from now. Excluding such unpleasant natural or human-caused catastrophes, without an extraterrestrial frontier, a planetary society will simply stagnate due to isolation, while other intelligent alien civilizations (as may exist) flourish and populate the galaxy.

Replicating robot system technology offers the human race very interesting options for the spread of life beyond the boundaries of Earth. Future generations of human beings might decide to create autonomous, self-replicating robot probes (von Neumann probes) and send these systems across the interstellar void on missions of exploration. Future generations of humans otherwise could elect to develop a closely knit (symbiotic) human-machine system—a highly automated interstellar ark—that is capable of crossing interstellar regions and then replicating itself when it encounters star systems with suitable planets and resources.

According to some scientists, any intelligent civilization that desires to explore a portion of the galaxy more than 100 light-years from their parent star would probably find it more efficient to use self-replicating robot probes. This galactic exploration strategy would produce the largest amount of directly sampled data about other star systems for a given period of exploration. One estimate suggests that the entire galaxy could be explored in about one million years, assuming the replicating interstellar probes could achieve speeds of at least one-tenth the speed of light. If other alien civilizations (should such exist) follow this approach, then the most probable initial contact between extraterrestrial civilizations would involve a self-replicating robot probe from one civilization encountering a self-replicating probe from another civilization.

If these encounters are friendly, the probes could exchange a wealth of information about their respective parent civilizations and any other civilizations previously encountered in their journeys through the galaxy. The closest terrestrial analogy would be a message placed in a very smart bottle that is then tossed into the ocean. If the smart bottle encounters another smart bottle, the two bump gently and provide each other a copy of their entire content of messages. One day, a beachcomber finds a smart bottle and discovers the entire collection of messages from across the world's oceans that has accumulated within.

If the interstellar probes have a hostile, belligerent encounter, they will most likely severely damage or destroy each other. In this case, the journey through the galaxy ceases for both probes and the wealth of information about alien civilizations, existent or extinct, vanishes. Returning to the simple-message-in-smart-bottle analogy here on Earth, a hostile encounter damages both bottles, they sink to the bottom of the ocean, and their respective information contents are lost forever. No beachcomber will ever discover either bottle and so will never have the chance of reading the interesting messages contained within.

One very distinct advantage of using interstellar robot probes in the search for other intelligent civilizations is the fact that these probes could also serve as a cosmic safety deposit box, carrying information about the technical, social, and cultural aspects of a particular civilization through the galaxy long after the parent civilization has vanished. The gold-anodized records that NASA engineers included on the *Voyager 1* and *2* spacecraft and the special plaques they placed on the *Pioneer 10* and *11* spacecraft are humans' first attempts at achieving a tiny degree of cultural immortality in the cosmos. (Chapter 9 discusses these spacecraft and the special messages they carry.)

Star-faring, self-replicating machines should be able to keep themselves running for a long time. One speculative estimate by exobiologists suggests that there may exist at present only 10 percent of all alien civilizations that ever arose in the Milky Way galaxy, the other 90 percent having perished. If this estimate is correct, then—on a simple statistical basis—nine out of every 10 robotic star probes within the galaxy could be the only surviving artifacts from long-dead civilizations. These self-replicating star probes would serve as emissaries across interstellar space and through eons of time. Here on Earth, the discovery and excavation of ancient tombs and other archaeological sites provides a similar contact through time with long-vanished peoples.

Perhaps later this century, human space explorers and/or their machine surrogates will discover a derelict alien robot probe or will recover an artifact the origins of which are clearly not from Earth. If terrestrial scientists and cryptologists are able to decipher any language or message contained on the derelict probe (or recovered artifact), humans may eventually learn about at least one other ancient alien society. The discovery of a functioning or derelict robot probe from an extinct alien civilization may also lead human investigators to many other alien societies. In a sense, by encountering and interrogating successfully an alien robot star probe, the human team of investigators may actually be treated to a delightful edition of the proverbial *Encyclopedia Galactica*—a literal compendium of the technical, cultural, and social heritage of thousands of extraterrestrial civilizations within the galaxy (most of which are probably now extinct). (Chapter 11 also addresses the issue of alien contact.)

There are a number of interesting ethical questions concerning the use of interstellar self-replicating probes. Is it morally right, or even equitable, for a self-replicating machine to enter an alien star system and harvest a portion of that star system's mass and energy to satisfy its own mission objectives? Does an intelligent species legally "own" its parent star, home planet, and any material or energy resources residing on other celestial

ANNAPOLIS MIDDLE SCHO
MEDIA CENTER

objects within its star system? Does it make a difference whether the star system is inhabited by intelligent beings? Or is there some lower threshold of galactic intelligence quotient (GIQ) below which star-faring races may ethically (on their own value scales) invade an alien star system and appropriate the resources that are needed to continue on their mission through the galaxy? If an alien robot probe enters a star system to extract resources, by what criteria does the smart machine judge the intelligence level of any indigenous life-forms? Should this smart robot probe avoid severely disturbing or contaminating existing life-bearing ecospheres?

Further discussion about and speculative responses to such intriguing SRS-related questions extends far beyond the scope of this chapter. However, the brief line of inquiry introduced here cannot end without at least mention of the most important question in cosmic ethics: Now that the human species has developed space technology, are humans and their solar system above (or below) any galactic appropriations threshold?

Close Encounters: The Consequences of Contact

Writers and filmmakers have thoroughly exploited the concept of contact with extraterrestrial life and the potential consequences for the human race. The publication of *The War of the Worlds* by H. G. Wells at the end of the 19th century started the very popular alien-invasion subgenre in science fiction.

This chapter explores the currently hypothetical subject of alien contact within the framework of two basic issues. The first issue involves the *UFO hypothesis*—the persistent notion that unidentified flying objects are under the control of extraterrestrial beings who are using these elusive spacecraft to survey and visit the Earth. The second issue involves the potential consequences (usually dire) of making contact with an alien life-form—especially one that has superior intelligence or technology. Selected examples from the science-fiction literature or the motion picture industry are used in the chapter to emphasize some recurring themes and potentially key points.

The classic science-fiction film *The Day the Earth Stood Still* appeared in theaters in 1951. Based on the short story *Farewell to the Master* by Harry Bates, director Robert Wise created this trendsetting alien-contact movie. The story opens when a flying saucer lands in a park in Washington, D.C., and out steps Klaatu (played by Michael Rennie)—a humanoid spokesperson from an advanced extraterrestrial civilization—and his powerful robot companion, Gort. As a historic note, this movie was the first major science-fiction film in which powerful aliens arrive not to conquer Earth but to help the human race decide to abandon warfare and join with other peaceful alien societies. Klaatu's message to the human race can be summarized quite simply: End the senseless nuclear arms race *before* attempting to travel out into space, or suffer lethal consequences from the powerful race of robots to which Gort belongs. Klaatu's message emphasizes that there is just no place in the "civilized" galaxy for an aggressive species like the human race. Due to humankind's development of nuclear energy and rocketry, Earth is now being regarded as a threat to the peaceful galactic community.

The film's plot has also been interpreted to contain religious symbolism, since Klaatu is killed in attempting to perform his peacekeeping mission and then is brought back to life by Gort. More apparent in the original short story by Bates but left quite ambiguous to the viewers of Wise's film is the interesting question of which alien, Klaatu the humanoid or Gort the robot, is really the powerful master.

In 1953, Sir Arthur C. Clarke published an enduring science-fiction novel entitled *Childhood's End*. His story starts with the apparently benign arrival of a powerful group of aliens called the Overlords. Operating from their giant spaceships in orbit around Earth, they help the human race usher in a golden age of peace and prosperity around the planet. The only price that people pay for this utopian era is the gradual loss of personal creativity and freedom. The end of Clarke's story involves an almost mystical twist: As a result of contact with the Overlords (who are eventually described as having a demonic "horns-and-tail" appearance), the human race loses its childhood and transforms to another level of existence.

Director Steven Spielberg produced two legendary alien-encounter movies that introduced the possibility of truly benign alien contact to millions of people around the world. In his 1977 science-fiction movie *Close Encounters of the Third Kind*, Spielberg addressed the issue of unidentified flying objects (UFOs). An apparently very friendly yet powerful alien race had selected certain human beings for a special rendezvous and instructed these people to gather at Devil's Tower in Wyoming. To provide some additional tension in the story, Spielberg has the U.S. Army sealing off the area and trying to prevent these "invited" people from entering the location. The movie climaxes with a spectacular encounter scene. An enormous alien mother ship arrives, and the first communication between the human race and the intelligent alien race (depicted by Spielberg as small, thin, gray beings with somewhat enlarged bald heads) takes place using musical notes. (Several movie reviewers have suggested that Spielberg wanted to make his aliens match some of the [unsubstantiated] alien abduction accounts that were popular in the 1950s and 1960s.) At the end of the film, the specially chosen humans, who successfully reached Devil's Tower, are invited by the aliens to board their ship and visit the stars. Within the language of UFO encounters, this was a benign, friendly encounter of the third kind.

The second Spielberg alien-encounter film, *E.T. the Extra-Terrestrial*, originally appeared in theaters in 1982 and was rereleased in 1985 and 2002. It is a classic "feel-good" story about a young Earth boy (named Elliot) and his delightful alien companion, nicknamed "E.T." As the movie opens, the friendly little alien arrives on Earth along with his companions. Working as extraterrestrial botanists, they explore a thickly wooded area in California and set about collecting plant specimens. However, little E.T. becomes stranded and panics when (human) government officials suddenly show up and try to capture some of the aliens. The remainder of Spielberg's delightful, Oscar-winning movie follows the adventures of E.T. and Elliot as they evade capture by the relentless government officials. This movie involves alien contact—but with a slightly different spin. Now, the humans are aggressively pursuing the alien. Ostensibly acting in the name of science, the government officials want to capture and possibly dissect (for research purposes) the obviously harmless but intelligent extraterrestrial creature who has been stranded here on Earth by accident. In one of the film's most charming scenes, E.T. manages to construct a special signaling device and contacts ("phones home") his companions, who are some distance away aboard their spaceship. Spielberg gives this alien-encounter movie an ending that is Hollywood entertainment magic at its very best.

This movie also raises a very interesting "galaxywide" morale issue, concerning the right of any intelligent species to collect living specimens of other life-forms that possess some obvious level of intelligence—even if the collection is being done in the name of science. In other words, do advanced alien races (*should such exist*) maintain zoos, laboratories, or research facilities that are populated with "less intelligent" living species that have been collected from other planets within their parent solar system or perhaps from life-supporting planets that were encountered in other star systems?

✧ Unidentified Flying Object (UFO)

The unidentified flying object (UFO) is a flying object (apparently) that is seen in the terrestrial skies by an observer who cannot determine its nature. The vast majority of such UFO sightings can, in fact, be explained by known phenomena. However, these phenomena may be beyond the knowledge or experience of the person making the observation. Common phenomena that have given rise to UFO reports include artificial Earth satellites, aircraft, high-altitude weather balloons, certain types of clouds, and even the planet Venus.

There are, nonetheless, some reported sightings that cannot be fully explained on the basis of the data available (which may be insufficient or too scientifically unreliable) or on the basis of comparison with known phenomena. It is the investigation of these relatively few UFO-sighting cases that has given rise, since the end of World War II, to the *UFO hypothesis,* a popular (though technically unfounded) hypothesis that speculates

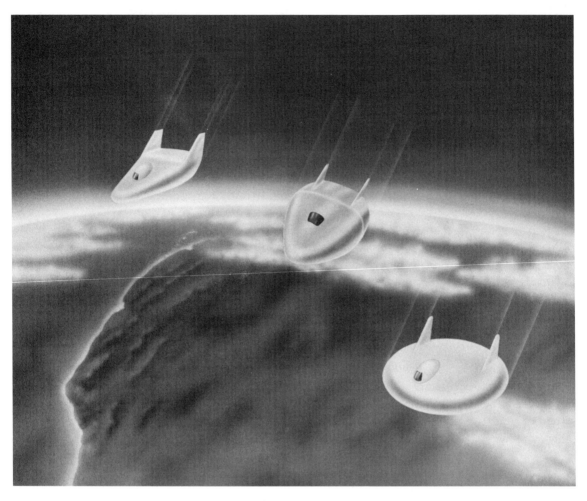

This artist's rendering depicts three lenticular bodies (from left to right: Ames M2–F1, Ames M1–L half-cone, and Langley design) explored by NASA in the 1960s as a possible means of landing an aerospace vehicle horizontally after atmospheric reentry. The lenticular shape of such wingless, lifting-body experimental aircraft was often likened to that of a "flying saucer" or a UFO. *(NASA/Dryden Flight Research Center)*

that these unidentified flying objects are under the control of extraterrestrial beings who are surveying and visiting Earth.

Modern interest in UFOs appears to have begun with a sighting report made by a private pilot named Kenneth Arnold. In June 1947, he reported seeing a mysterious formation of shining disks in the daytime near Mount Rainier in the state of Washington. When newspaper reporters heard of his account of "shining saucerlike disks," the popular term *flying saucer* was born.

In 1948, the U.S. Air Force (USAF) began to investigate these UFO reports. Project Sign was the name given by the U.S. Air Force to its initial study of UFO phenomena. In the late 1940s, Project Sign was replaced by Project Grudge, which in turn became the more familiar Project Blue Book. Under Project Blue Book, the U.S. Air Force investigated many UFO reports from 1952 to 1969. Then, on December 17, 1969, the secretary of the USAF announced the termination of Project Blue Book.

The USAF decision to discontinue UFO investigations was based on the following circumstances: (1) an evaluation of a report prepared by the University of Colorado and entitled, "Scientific Study of Unidentified Flying Objects" (this report is also often called the Condon report after its principal author, Edward Uhler Condon [1902–74]); (2) a review of this University of Colorado report by the National Academy of Sciences; (3) previous UFO studies; and (4) U.S. Air Force experience from nearly two decades of UFO report investigations.

As a result of these investigations and studies and of experience gained from UFO reports since 1948, the conclusions of the U.S. Air Force were: (1) No UFO reported, investigated, and evaluated by the USAF ever gave any indication of threatening national security; (2) there was no evidence submitted to or discovered by the USAF that sightings that had been categorized as "unidentified" represent technological developments or principles beyond the range of present-day scientific knowledge; and (3) there

This unusual postage stamp, issued by Grenada in 1978, commemorates research into unidentified flying objects, or UFOs. The design features a flying-saucer illustration (on the left) and a 1965 photograph of an unexplained object or phenomenon (on the right). *(Author)*

was no evidence to indicate that the sightings categorized as "unidentified" are extraterrestrial vehicles.

With the termination of Project Blue Book, the U.S. Air Force regulation establishing and controlling the program for investigating and analyzing UFOs was rescinded. All documentation regarding Project Blue Book investigations was then transferred to the Modern Military Branch, National Archives and Records Service, 8th Street and Pennsylvania Avenue N.W., Washington, D.C. 20408. This material is presently available for public review and analysis. If a person wishes to review these files, they need simply to obtain a researcher's permit from the National Archives and Record Service. Of a total of 12,618 sightings reported to Project Blue Book, 701 remained "unidentified" when the USAF ended the project. Since the termination of Project Blue Book, nothing has occurred that has caused the USAF or any other federal agency to support a resumption of UFO investigations. Today, reports of unidentified objects entering North American air space are still of interest to the military—primarily as part of its overall defense surveillance program and support of heightened national anti-terrorism activities. But beyond those national defense-related missions, the U.S. Air Force no longer investigates reports of UFO sightings.

During the past half-century, the subject of UFOs has evoked strong opinions and emotions. For some people, the belief in or study of UFOs has assumed almost the dimensions of a religious quest. Other individuals remain nonbelievers or at least very skeptical concerning the existence of alien beings and elusive vehicles that never quite seem to manifest themselves to scientific investigators or to competent government authorities. Regardless of one's conviction, nowhere has the debate about UFOs in the United States been more spirited than over the events that unfolded near the city of Roswell, New Mexico, in summer 1947. This event, popularly known as the Roswell incident, has become widely celebrated as a UFO encounter. Numerous witnesses, including former military personnel and respectable members of the local community, have come forward with stories of humanoid beings, alien technologies, and government cover-ups that have caused even the most skeptical observer to pause to consider the reported circumstances. Inevitably, over the years, these tales have spawned countless articles, books, and motion pictures concerning visitors from outer space who crashed in the New Mexico desert.

As a result of increasing interest and political pressure concerning the Roswell incident, in February 1994, the U.S. Air Force was informed that the General Accounting Office (GAO), an investigative agency of Congress, planned to conduct a formal audit to determine the facts regarding the reported crash of a UFO in 1947 at Roswell, New Mexico. The GAO's investigative task actually involved numerous federal agencies, but the focus was on the U.S. Air Force—the agency most often accused of hiding information and records concerning the Roswell incident. The GAO research team conducted an extensive search of U.S. Air Force archives, record centers, and scientific facilities. Seeking information that might help explain peculiar tales of odd wreckage and alien bodies, the researchers reviewed a large number of documents that concerned a variety of events including aircraft crashes, errant missile tests (from White Sands, New Mexico), and nuclear mishaps.

This extensive research effort revealed that the Roswell incident was not even considered a UFO event until the 1978–80 time frame. Prior to that, the incident was generally dismissed because officials in the U.S. Army Air Force (predecessor to the U.S. Air Force) had originally identified the debris recovered as being that of a weather balloon. The GAO research effort located no records at existing air-force offices that indicated any cover-up by the USAF or that provided any indication of the recovery of an alien spacecraft or its occupants. However, records were located and investigated concerning a then top-secret balloon project, Project Mogul, which attempted to monitor Soviet nuclear tests. Comparison of all information that was developed or obtained during this effort indicated that the material recovered near Roswell was consistent with a balloon device and most likely from one of the Project Mogul balloons that had not been recovered previously. This government response to contemporary inquiries concerning the Roswell incident is described in an extensive report that was released in 1995 by Headquarters USAF and entitled: "The Roswell Report: Fact versus Fiction in the New Mexico Desert." While the National Aeronautics and Space Administration (NASA) is the current federal agency focal point for answering public inquiries to the White House concerning UFOs, the civilian space agency is not engaged in any research program involving these UFO phenomena or sightings—nor is any other agency of the U.S. government.

One interesting result that emerged from Project Blue Book is a popular scheme, developed by Dr. J. Allen Hynek (1910–60), to classify or categorize UFO sighting reports. Six classification levels were used to organize these UFO reports. A type-A UFO report generally involved seeing bright lights in the night sky. These sightings usually turn out to be a planet (typically Venus), a satellite, an airplane, or meteors. A type-B UFO report often involved the daytime observation of shining disks (that is, flying saucers) or cigar-shaped metal objects. This type of sighting usually ended up as a weather balloon, a blimp or lighter-than-air ship, or even a deliberate prank or hoax. A type-C UFO report involved unknown images appearing on a radar screen. These signatures might linger, be tracked for a few moments, or simply appear and then quickly disappear—often to the amazement and frustration of the scope operator. Such radar visuals frequently turn out to be something like swarms of insects, flocks of birds, unannounced aircraft, and perhaps the unusual phenomena that radar operators like to call "angels." To radar operators, angels are anomalous with radar-wave propagation phenomena.

Close encounters of the first kind (visual sighting of a UFO at moderate to close range) represent the type-D UFO reports. Typically, the observer reports something unusual in the sky that "resembles an alien spacecraft." In the type-E UFO report, not only does the observer claim to have seen the alien spaceship but also reports the discovery of some physical evidence in the terrestrial biosphere (such as scorched ground, radioactivity, mutilated animals, etc.) that is associated with the alien craft's visit. This type of sighting has been named a close encounter of the second kind. Finally, in the type F UFO report, which is also called a close encounter of the third kind, the observer claims to have seen and sometimes to have been contacted by the alien visitors. Extraterrestrial contact stories range from simple sightings of "UFOnauts" to communication with them (usually telepathic) to cases of kidnapping and then release of the terrestrial observer. There are even some reported stories in which a terran (that is, person from Earth) was kidnapped and then seduced by an alien visitor—a challenging task of romantic compatibility even for an advanced star-faring species!

Despite numerous stories about such UFO encounters, not a single shred of scientifically credible, indisputable evidence has yet to be acquired. If scientists are asked to judge these reports, some arbitrary proof scale is needed as a guide to help them determine what type of data or testimony is actually necessary to provide convincing evidence that the "little green men" (LGM) have indeed arrived in their flying saucer.

UFO Report Classifications

The follow classification scheme for UFO reports was developed by Dr. J. Allen Hynek and USAF Project Blue Book.

A. Nocturnal (nighttime) light
B. Diurnal (daytime disk)
C. Radar contact (radar visual [RV])
D. Visual sighting of alien craft at modest to close range (also called close encounter of the first kind [CE I])
E. Visual sighting of alien craft plus discovery of (hard) physical evidence of craft's interaction with terrestrial environment (also called close encounter of the second kind [CE II])
F. Visual sighting of aliens themselves, including possible physical contact (also called close encounter of the third kind [CE III])

PROPOSED "PROOF SCALE" TO ESTABLISH THE EXISTENCE OF UFOs

The following arbitrary "proof scale" has been constructed from the work of Dr. J. Allen Hynek and the USAF Project Blue Book.

Highest Value*

(1) The alien visitors themselves or the alien spaceship

(2) Irrefutable physical evidence of a visit by aliens or the passage of their spaceship

(3) Indisputable photograph of an alien spacecraft or one of its occupants

(4) Human eyewitness reports

Lowest Value

(*from standpoint of the scientific method and validation of the UFO hypothesis with "hard" technical data)

Unfortunately, scientists do not have any convincing data to support UFO "proof-scale" categories 1 to 3. Instead, all the information available involve large quantities of proof-scale category-four eyewitness accounts of various UFO encounters. Even the most sincere human testimony changes in time and is often subject to wide variations and contradiction. The scientific method puts very little weight on human testimony in validating a hypothesis.

Even from a more philosophical point of view, it is very difficult to accept the UFO hypothesis logically. Although intelligent life may certainly have evolved elsewhere in the universe, the UFO encounters reported to date hardly reflect the logical exploration patterns and encounter sequences that scientists might anticipate from an advanced, star-faring alien civilization.

In terms of humans' current understanding of the laws of physics, interstellar travel appears to be an extremely challenging, if not technically impossible, undertaking. Any alien race that developed the advanced technologies that are necessary to travel across vast interstellar distances would most certainly be capable of developing sophisticated remote-sensing technologies. With these remote-sensing technologies, they could study the Earth essentially undetected—unless, of course, they wanted to be detected. And if they wanted to make contact, they most surely could observe where the Earth's population centers are and land in places where they could communicate with competent authorities. It is insulting not only to their intelligence but to our own human intelligence as well to think that these alien visitors would repeatedly contact only people who are in remote, isolated areas, scare the dickens out of them, and then lift off into the sky. Why not once land in the middle of the Rose Bowl during a football game or near the site of an international meeting of astronomers and astrophysicists? And why conduct only short momentary intrusions into the terrestrial biosphere? After all, the Viking landers that NASA sent to Mars gathered data for years. It is hard to imagine that an advanced culture would make the tremendous resource investment to send a robot probe or even to arrive here themselves and then only flicker through an encounter with just a very few beings on this planet. Are most human beings that uninteresting? If that is the case, then why so many reported visits? From a simple exercise of logic, the UFO hypothesis just does not make sense—terrestrial or extraterrestrial!

Hundreds of UFO reports have been made since the late 1940s. Again, why is Earth so interesting? Is the planet at a galactic crossroads? Are the outer planets of the solar system an "interstellar truck stop" where alien starships pull in and refuel? (Some people have already proposed this hypothesis.) Let us play a simplified interstellar traveler game to see if so many reported visits are realistic, even if Earth and human beings are very interesting on some cosmic scale. First, we assume that the Milky Way galaxy of more than 100 billion stars contains about 100,000 different starfaring alien civilizations that are more or less uniformly dispersed. (This is a very optimistic number according to both the Drake equation and scientists who have speculated about the likelihood of Kardashev Type II civilizations.) Then each of these extraterrestrial civilizations has, in principle, one million other star systems to visit without interfering with any other civilization. (Yes, the Milky Way galaxy is a really big place!) What are the odds of two of these civilizations both visiting humans' solar system and each only casually exploring planet Earth during the last five decades? The only logical conclusion that can be drawn is that the UFO-encounter reports are not credible indications of extraterrestrial visitations. Yet, despite the lack of scientific evidence, UFO-related Web sites are among the most popular and frequently visited on the Internet.

✧ Consequences of Interstellar Contact

Starting in the mid-1990s, the discovery of planets around other stars has renewed scientific speculation about the possibility of intelligent life elsewhere in the Milky Way galaxy. Should such intelligent life exist, would they also be interested in searching for and contacting other intelligent beings like us? Just what would happen if human beings contact an extraterrestrial civilization? No one on Earth can really say for sure. However, this contact will very probably be one of the most momentous events in all human history.

The postulated interstellar contact can be direct or indirect. Direct contact might involve a visit to Earth by a starship from an advanced stellar

This interesting 1992 painting by Pat Rawlings is entitled *The Key*. It depicts the hypothetical but incredibly exciting moment in the future when a human space explorer serendipitously discovers an alien artifact on Mars (or perhaps elsewhere in the solar system). As many advocates of space exploration proclaim, the human drive to explore the unknown, spawned by the primordial biological instinct to survive, will lead future generations to unimaginable adventures and new discoveries beyond the boundaries of Earth. Space technology opens up unlimited new frontiers for social and technical growth throughout the solar system and eventually among the stars. *(© Pat Rawlings, the artist; used here with permission)*

civilization. It could also take the form of the discovery of an alien probe, artifact, or derelict spaceship in the outer regions of our solar system. Some space travel experts have suggested that the hydrogen- and helium-rich giant outer planets might serve as convenient "fueling stations" for passing interstellar spaceships from other worlds. Indirect contact via radio frequency communication represents a more probable contact pathway (at least from a contemporary terrestrial viewpoint). The consequences of a successful search for extraterrestrial intelligence (SETI) would be nothing short of extraordinary. Were scientists able to locate and identify but a single extraterrestrial signal, humankind would know immediately one great truth: We are not alone. We might also learn that the universe is teeming with life.

The overall impact of this postulated interstellar contact will depend on the circumstances surrounding the event. If it happens by accident or after only a few years of searching, the news, once verified, would surely startle most citizens of this world. If, however, intelligent alien signals were detected only after an extended effort, lasting generations and involving extensive search facilities, the terrestrial impact of the event might be less overwhelming.

The reception and decoding of a radio-frequency, optical, or other type signal from an extraterrestrial civilization in the depths of space offers the promise of practical and philosophical benefits for all humanity. Response to that signal, however, involves a potential planetary risk. If scientists ever intercept and validate an alien signal, then humans (as a planetary society) can decide to respond. The majority of people may also choose not to respond. If we are suspicious of the motives of the alien culture that sent the message, we are under no obligation to answer. There would be no practical way for them to realize that their signal had been intercepted, decoded, and understood by the intelligent inhabitants of a tiny world called Earth.

Optimists emphasize the friendly nature of such an interstellar contact and anticipate large technical gains for our planetary society, including the reception of information and knowledge of extraordinary value.

EXOTHEOLOGY

Exotheology is the organized body of opinions concerning the impact that space exploration and the possible discovery of life beyond the boundaries of Earth would have on contemporary terrestrial religions. On Earth, theology involves the study of the nature of God and the relationship of human beings to God.

Throughout human history, people have gazed into the night sky and pondered the nature of God. They searched for those basic religious truths and moral beliefs that define how an individual should interact with his or her Creator and toward each other. The theologies found within certain societies (especially ancient ones) sometimes involved a collection of gods. For people in such societies, the plurality of specialized gods proved useful in explaining natural phenomena

as well as in codifying human behavior. For example, in the pantheism of ancient Greece, an act that really displeased Zeus—the lord of the gods who ruled Earth from Mount Olympus—would cause him to hurl a powerful thunderbolt at the offending human. Thus, the fear of "getting zapped by Zeus" often helped ancient Greek societies maintain a set of moral standards. Other terrestrial religions stressed belief in a single God. Judaism, Christianity, and Islam represent our planet's three major monotheistic religions. Each has a collection of dogma and beliefs that define acceptable moral behavior by human beings and also identify individual rewards (for example, personal salvation) and punishments (for example, eternal damnation) for adherence or transgressions, respectively.

They imagine that there will be numerous scientific and technological benefits from such contacts. However, because of the long round-trip times that are associated with speed-of-light-limited interstellar communications (perhaps decades or centuries), any information exchange will most likely be in the form of semi-independent transmissions. Each transmission burst would contain a bundle of significant facts about the sending society—such as planetary data, its life-forms, its age, its history, its philosophies and beliefs, and whether it has contacted other alien cultures successfully. An interstellar dialogue with questions asked and answered in somewhat rapid succession would not be as practical. Therefore, during the period of a century or more, the people of Earth might receive a wealth of information at a gradual enough rate so as to assemble a comprehensive picture of the alien civilization without inducing severe culture shock here at home.

Some scientists feel that if we establish interstellar contact successfully, we probably will not be the first planetary civilization to have accomplished this feat. In fact, they speculate that interstellar communications may have been going on since the first intelligent civilizations evolved in the Milky Way galaxy about four or five billion years ago. One of the most exciting consequences of this type of interstellar conversation would be the accumulation by all participants of an enormous body of

In the 21st century, advanced space exploration missions could lead to the discovery of simple alien life-forms on other worlds within the solar system. This important scientific achievement would undoubtedly rekindle serious interest in one of the oldest philosophical questions that has puzzled a great number of people throughout history: Are human beings the only intelligent species in the universe? If not, what happens if contact is made with an alien intelligence? What might be our philosophical and theological relationship with such intelligent creatures?

Some space age theologians are now starting to grapple with these intriguing questions and many similar ones. For example, *should it exist,* would an alien civilization that is much older than our planetary civilization have a sig-nificantly better understanding of the universe and, by extrapolation, a clearer understanding of the nature of God? Here on Earth, many brilliant scientists, such as Sir Isaac Newton and Albert Einstein, regarded their deeper personal understanding of the physical universe as an expanded perception of its Creator. Would these (hypothetical) advanced alien creatures be willing to share their deeper understanding of God? And if they do decide to share that deeper insight into the divine nature, what impact would that communication have on terrestrial religions? Exotheology involves such interesting speculations as well as the extrapolation of teachings found within terrestrial religions to accommodate the expanded understanding of the universe that will be brought about by the scientific discoveries of space exploration and modern astronomy.

information and knowledge that has been passed down from alien race to alien race since the beginning of the galaxy's communicative phase. Included in this vast body of knowledge—something we might call the "galactic heritage"—could be the entire natural and social history of numerous species and planetary civilizations. Also included would be an extensive compendium of astrophysical data that extend back countless millennia. Access to such a collection of scientific data would allow the best and brightest minds on Earth to develop accurate new insights into the origin and destiny of the universe.

However, interstellar contact should lead to far more than merely an exchange of scientific knowledge. Humankind would discover other social forms and structures, probably better capable of self-preservation and genetic evolution. We would also discover new forms of beauty and become aware of different endeavors that promote richer, more rewarding lives. Such contacts might also lead to the development of branches of art or science that simply cannot be undertaken by just one planetary civilization. Rather they would require joint, multiple-civilization participation across interstellar distances. Most significant, perhaps, is the fact that interstellar contact and communication would represent the end of the cultural isolation of the human race. The contact optimists further speculate that our own civilization would be invited to enter a sophisticated "cosmic community" as mature "adults" who are proud of our own human heritage—rather than remaining isolated with a destructive tendency to annihilate each other in childish planetary rivalries. Indeed, perhaps the very survival of the human race ultimately depends on finding ourselves cast in a larger cosmic role—a role that is far greater in significance than any human being can now imagine.

We should also speculate about the possible risks that could accompany confirming our existence to an alien culture that is most likely far more advanced and powerful than our own. Contact pessimists suggest that such risks range from planetary annihilation to the humiliation of the human race. For discussion purposes, these risks can be divided into four general categories: invasion, exploitation, subversion, and culture shock.

As mentioned previously, the invasion of Earth is a popular and recurring theme in science fiction. By sending out signals into the cosmic void actively or by responding to intelligent signals that we have detected and decoded, the human race would be revealing its existence and announcing the fact that Earth is a habitable planet. Within this contact scenario, Earth might be invaded by vastly superior beings who are set on conquering the galaxy. Another major interstellar contact hazard is exploitation. Human beings could appear to be a very primitive form of conscious life—perhaps the level of an experimental animal or an unusual pet.

Another interstellar contact hazard is that of subversion. This appears to be a more plausible and subtle form of contact risk because it can occur with an exchange of signals. Here, an advanced alien race—under the guise of teaching and helping the human race join a cosmic community—might actually trick us into building devices that would allow "Them" to conquer "Us." The alien civilization would not necessarily have to make direct contact since their interstellar "Trojan horse" might arrive via radio frequency signals. Since computer worms, viruses, and Trojan horses are a constant Internet threat here on Earth, why not a similar, but more sophisticated, interstellar contact threat?

The last general contact risk is massive culture shock. Some individuals have expressed concerns that even mere contact with a vastly superior extraterrestrial race could prove extremely damaging to human psyches, despite the best intentions of the alien race. Terrestrial cultures, philosophies, and religions that now place human beings at the very center of creation would have to be "modified and expanded" to compensate for the confirmed existence of other, far superior intelligent beings. We would now have to "share the universe" with someone or something better and more powerful than we are. As the dominant species on this planet, we must seriously consider whether the majority of human beings could accept this new role.

Advances in space technology are encouraging many people to consider the possibility of the existence of intelligent species elsewhere in the universe. As this happens, we must also keep asking ourselves a more fundamental question: Are human beings generally prepared for the positive identification of such a fact? Will contact with intelligent aliens open up a golden age on Earth or initiate devastating cultural regression?

But the choice of initiating interstellar contact may no longer really be ours. In addition to the radio and television broadcasts that are leaking out into the galaxy at the speed of light, the powerful radio/radar telescope at the Arecibo Observatory was utilized to beam an interstellar message of friendship to the fringes of the Milky Way galaxy on November 16, 1974. We have, therefore, already announced our presence to the galaxy and should not be too surprised if someone or something eventually answers.

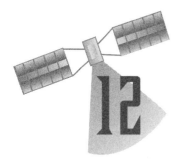

Conclusion

....

This book has provided a space age examination of the fundamental question: Is life, especially intelligent life, unique to Earth? Human interest in the origins of life and the possibility of life on other worlds extends back deep into antiquity. Today, as a result of space technology, the scope of those early perceptions has expanded well beyond the reaches of this solar system—to other stars of the Milky Way galaxy, to the vast interstellar clouds that serve as stellar nurseries, and beyond to numerous galaxies that populate the seemingly infinite expanse of outer space.

Astronomical evidence now suggests that planet formation is a natural part of stellar evolution. If life originates on "suitable" planets whenever it can (as many exobiologists currently hypothesize), then knowing how abundant such suitable planets are in the Milky Way galaxy would allow scientists here on Earth to make more credible guesses about where to search for extraterrestrial intelligence and what the basic chances are of finding intelligent life beyond this solar system.

One of the fundamental characteristics of human nature is our desire to communicate. In recent years, we have begun to respond to a deep cosmic yearning to reach beyond this solar system to other star systems—hoping not only that someone or something is out there but that "they" will eventually "hear us" and perhaps even return our message.

Because of the vast distances between even nearby stars, when scientists say "interstellar communication," they are not talking about communication in "real time." On Earth, radio-wave communication does not involve a very perceptible time lag—that is, messages and responses are normally received immediately after they are sent. In contrast, electromagnetic waves take many light-years to cross the interstellar void from star system to star system, so our initial attempts at interstellar communication actually have been more like putting a message in a bottle and tossing it into the "cosmic sea" or else placing a message in a type of interstellar time capsule for some future generation of human or alien beings to discover.

Positive, active attempts to communicate with any alien civilizations that might exist among the stars are often referred to as CETI—an acronym meaning communication with extraterrestrial intelligence. If, on the other hand, humankind passively scans the skies for signs of an advanced extraterrestrial civilization or patiently listens for their radio-wave messages, the process is often referred to as SETI—meaning the search for extraterrestrial intelligence. While the acronyms appear quite similar, the societal implications are quite different.

Starting in 1960, there have been several serious SETI efforts, the vast majority of which involved listening to selected portions of the microwave spectrum in hopes of detecting structured electromagnetic signals that were indicative of the existence of some intelligent extraterrestrial civilization among the stars. To date, none of these efforts has provided any positive evidence that such "intelligent" (that is, coherent and artificially produced versus naturally occurring) radio signals exist. However, SETI researchers have only examined a relatively small portion of the billions of stars in the Milky Way galaxy. Furthermore, they have generally only listened to rather narrow portions of the electromagnetic spectrum. Optimistic investigators, therefore, suggest that "the absence of evidence is not necessarily the evidence of absence." But detractors counter by suggesting that SETI is really an activity without a subject. Such detractors also regard the SETI researcher as a modern Don Quixote who is tilting with cosmic windmills.

But the Milky Way is a vast place, and it is only within the past few decades that the people of Earth have enjoyed radio, television, and other information technologies that are sophisticated enough to cross the threshold of some minimal interstellar communication technology horizons. A little more than a century ago, for example, Earth could have been "bombarded" quite literally with many alien radio-wave signals—but no one here had the technology capable of receiving or interpreting such hypothetical signals. After all, even the universe's most colossal natural signal, the remnant microwave background from the big bang was just detected and recognized in the mid-1960s. Yet, this interesting signal is detectable (as static) by any common television set that is connected to a receiving antenna.

Early in the space age, a few scientists did attempt boldly to communicate actively "over time" with alien civilizations. They did this by including carefully prepared messages on four robot spacecraft (*Pioneer 10* and *11* and *Voyager 1* and *2*) that were scheduled to travel beyond the solar system and also by transmitting *intentionally* a very powerful radio message to a special group of stars, using the world's largest radio telescope, the Arecibo

Observatory. These technical deeds may be viewed as the human race's inaugural attempts at CETI.

It may come as a bit of a surprise to most readers that the people of Earth have also been leaking radio frequency signals *unintentionally* into the galaxy since the middle of the 20th century. Imagine the impact that some of our early television shows could have on an alien civilization that is capable of intercepting and reconstructing the television signals. The earliest of these TV broadcast signals are now about 60 light-years into the galaxy. Who knows "what alien society" is possibly examining the *Ed Sullivan Show* during which Elvis Presley made his initial television appearance. Should these speculative circumstances occur, a group of alien scientists, philosophers, and religious leaders might right now be engaged in a very heated debate concerning the true significance of the special message from Earth that proclaims: "You Ain't Nothing But a Hound Dog!"

Perhaps even more interesting is the fact that at some point later in this century, a coherent, "artificial" signal could be received and deciphered during a scientifically based SETI effort. This (for now) hypothetical event would provide the human race with a very interesting opportunity to turn SETI into CETI. Do the people of Earth respond to this alien message? If so, what might we say, and who should speak for Earth?

Chronology

. . . .

✧ **ca. 3000 B.C.E. (to perhaps 1000 B.C.E.)**
Stonehenge erected on the Salisbury Plain of Southern England (possible use: ancient astronomical calendar for prediction of summer solstice)

✧ **ca. 1300 B.C.E.**
Egyptian astronomers recognize all the planets visible to the naked eye (Mercury, Venus, Mars, Jupiter, and Saturn), and they also identify more than 40 star patterns or constellations

✧ **ca. 500 B.C.E.**
Babylonians devise zodiac, which is later adopted and embellished by Greeks and used by other early peoples

✧ **ca. 375 B.C.E.**
The early Greek mathematician and astronomer Eudoxus of Cnidos starts codifying the ancient constellations from tales of Greek mythology

✧ **ca. 275 B.C.E.**
The Greek astronomer Aristarchus of Samos suggests an astronomical model of the universe (solar system) that anticipates the modern heliocentric theory proposed by Nicolaus Copernicus. However, these ideas, which Aristarchus presents in his work *On the Size and Distances of the Sun and the Moon*, are essentially ignored in favor of the geocentric model of the universe proposed by Eudoxus of Cnidus and endorsed by Aristotle

✧ **ca. 129 B.C.E.**
The Greek astronomer Hipparchus of Nicaea completes a catalog of 850 stars that remains important until the 17th century

✧ ca. 60 C.E.

The Greek engineer and mathematician Hero of Alexandria creates the aeoliphile, a toylike device that demonstrates the action-reaction principle that is the basis of operation of all rocket engines

✧ ca. 150 C.E.

Greek astronomer Ptolemy writes *Syntaxis* (later called the *Almagest* by Arab astronomers and scholars)—an important book that summarizes all the astronomical knowledge of the ancient astronomers, including the geocentric model of the universe that dominates Western science for more than one and a half millennia

✧ 820

Arab astronomers and mathematicians establish a school of astronomy in Baghdad and translate Ptolemy's work into Arabic, after which it became known as *al-Majisti* (The great work), or the *Almagest,* by medieval scholars

✧ 850

The Chinese begin to use gunpowder for festive fireworks, including a rocketlike device

✧ 1232

The Chinese army uses fire arrows (crude gunpowder rockets on long sticks) to repel Mongol invaders at the battle of Kaifung (Kuifeng). This is the first reported use of the rocket in warfare

✧ 1280–90

The Arab historian al-Hasan al-Rammah writes *The Book of Fighting on Horseback and War Strategies,* in which he gives instructions for making both gunpowder and rockets

✧ 1379

Rockets appear in western Europe; they are used in the siege of Chioggia (near Venice), Italy

✧ 1420

The Italian military engineer Joanes de Fontana writes *Book of War Machines,* a speculative work that suggests military applications of gunpowder rockets, including a rocket-propelled battering ram and a rocket-propelled torpedo

✧ 1429

The French army uses gunpowder rockets to defend the city of Orléans. During this period, arsenals throughout Europe begin to test various types of gunpowder rockets as an alternative to early cannons

✧ ca. 1500

According to early rocketry lore, a Chinese official named Wan-Hu attempted to use an innovative rocket-propelled kite assembly to fly through the air. As he sat in the pilot's chair, his servants lit the assembly's 47 gunpowder (black powder) rockets. Unfortunately, this early rocket test pilot disappeared in a bright flash and explosion

✧ 1543

The Polish church official and astronomer Nicolaus Copernicus changes history and initiates the Scientific Revolution with his book *De Revolutionibus Orbium Coelestium* (On the revolutions of the heavenly spheres). This important book, published while Copernicus lay on his deathbed, proposed a Sun-centered (heliocentric) model of the universe in contrast to the longstanding Earth-centered (geocentric) model advocated by Ptolemy and many of the early Greek astronomers

✧ 1608

The Dutch optician Hans Lippershey develops a crude telescope

✧ 1609

The German astronomer Johannes Kepler publishes *New Astronomy*, in which he modifies Nicolaus Copernicus's model of the universe by announcing that the planets have elliptical orbits rather than circular ones. Kepler's laws of planetary motion help put an end to more than 2,000 years of geocentric Greek astronomy

✧ 1610

On January 7, 1610, Galileo Galilei uses his telescope to gaze at Jupiter and discovers the giant planet's four major moons (Callisto, Europa, Io, and Ganymede). He proclaims this and other astronomical observations in his book, *Sidereus Nuncius* (Starry messenger). Discovery of these four Jovian moons encourages Galileo to advocate the heliocentric theory of Nicolaus Copernicus and brings him into direct conflict with church authorities

✧ 1642

Galileo Galilei dies while under house arrest near Florence, Italy, for his clashes with church authorities concerning the heliocentric theory of Nicolaus Copernicus

✧ 1647

The Polish-German astronomer Johannes Hevelius publishes *Seleno-graphia*, in which he provides a detailed description of features on the surface (near side) of the Moon

✧ 1680

Russian czar Peter the Great sets up a facility to manufacture rockets in Moscow. The facility later moves to St. Petersburg and provides the czarist army with a variety of gunpowder rockets for bombardment, signaling, and nocturnal battlefield illumination

✧ 1687

Financed and encouraged by Sir Edmond Halley, Sir Isaac Newton publishes his great work, *Philosophiae Naturalis Principia Mathematica* (Mathematical principles of natural philosophy). This book provides the mathematical foundations for understanding the motion of almost everything in the universe including the orbital motion of planets and the trajectories of rocket-propelled vehicles

✧ 1780s

The Indian ruler Hyder Ally (Ali) of Mysore creates a rocket corps within his army. Hyder's son, Tippo Sultan, successfully uses rockets against the British in a series of battles in India between 1782 and 1799

✧ 1804

Sir William Congreve writes *A Concise Account of the Origin and Progress of the Rocket System* and documents the British military's experience in India. He then starts the development of a series of British military (black-powder) rockets

✧ 1807

The British use about 25,000 of Sir William Congreve's improved military (black-powder) rockets to bombard Copenhagen, Denmark, during the Napoleonic Wars

✧ 1809

The brilliant German mathematician, astronomer, and physicist Carl Friedrich Gauss publishes a major work on celestial mechanics that revolutionizes the calculation of perturbations in planetary orbits. His work paves the way for other 19th-century astronomers to mathematically anticipate and then discover Neptune (in 1846), using perturbations in the orbit of Uranus

✧ 1812

British forces use Sir William Congreve's military rockets against American troops during the War of 1812. British rocket bombardment of Fort William McHenry inspires Francis Scott Key to add "the rocket's red glare" verse in the "Star Spangled Banner"

✧ 1865

The French science fiction writer Jules Verne publishes his famous story *De la terre a la lune* (From the Earth to the Moon). This story interests many people in the concept of space travel, including young readers who go on to become the founders of astronautics: Robert Hutchings Goddard, Hermann J. Oberth, and Konstantin Eduardovich Tsiolkovsky

✧ 1869

American clergyman and writer Edward Everett Hale publishes *The Brick Moon*—a story that is the first fictional account of a human-crewed space station

✧ 1877

While a staff member at the U.S. Naval Observatory in Washington, D.C., the American astronomer Asaph Hall discovers and names the two tiny Martian moons, Deimos and Phobos

✧ 1897

British author H. G. Wells writes the science-fiction story *The War of the Worlds*—the classic tale about extraterrestrial invaders from Mars

✧ 1903

The Russian technical visionary Konstantin Eduardovich Tsiolkovsky becomes the first person to link the rocket and space travel when he publishes *Exploration of Space with Reactive Devices*

✧ 1918

American physicist Robert Hutchings Goddard writes *The Ultimate Migration*—a far-reaching technology piece within which he postulates the use of an atomic-powered space ark to carry human beings away from a dying Sun. Fearing ridicule, however, Goddard hides the visionary manuscript and it remains unpublished until November 1972—many years after his death in 1945

✧ 1919

American rocket pioneer Robert Hutchings Goddard publishes the Smithsonian monograph *A Method of Reaching Extreme Altitudes*. This important work presents all the fundamental principles of modern rocketry. Unfortunately, members of the press completely miss the true significance of his technical contribution and decide to sensationalize his comments about possibly reaching the Moon with a small, rocket-propelled package. For such "wild fantasy," newspaper reporters dubbed Goddard with the unflattering title of "Moon man"

✦ 1923

Independent of Robert Hutchings Goddard and Konstantin Eduardovich Tsiolkovsky, the German space-travel visionary Hermann J. Oberth publishes the inspiring book *Die Rakete zu den Planetenräumen* (The rocket into planetary space)

✦ 1924

The German engineer Walter Hohmann writes *Die Erreichbarkeit der Himmelskörper* (The attainability of celestial bodies)—an important work that details the mathematical principles of rocket and spacecraft motion. He includes a description of the most efficient (that is, minimum energy) orbit transfer path between two coplanar orbits—a frequently used space operations maneuver now called the Hohmann transfer orbit

✦ 1926

On March 16 in a snow-covered farm field in Auburn, Massachusetts, American physicist Robert Hutchings Goddard makes space technology history by successfully firing the world's first liquid-propellant rocket. Although his primitive gasoline (fuel) and liquid oxygen (oxidizer) device burned for only two and one half seconds and landed about 60 meters away, it represents the technical ancestor of all modern liquid-propellant rocket engines.

In April, the first issue of *Amazing Stories* appears. The publication becomes the world's first magazine dedicated exclusively to science fiction. Through science fact and fiction, the modern rocket and space travel become firmly connected. As a result of this union, the visionary dream for many people in the 1930s (and beyond) becomes that of interplanetary travel

✦ 1929

German space-travel visionary Hermann J. Oberth writes the award-winning book *Wege zur Raumschiffahrt* (Roads to space travel) that helps popularize the notion of space travel among nontechnical audiences

✦ 1933

P. E. Cleator founds the British Interplanetary Society (BIS), which becomes one of the world's most respected space-travel advocacy organizations

✧ 1935

Konstantin Tsiolkovsky publishes his last book, *On the Moon*, in which he strongly advocates the spaceship as the means of lunar and interplanetary travel

✧ 1936

P. E. Cleator, founder of the British Interplanetary Society, writes *Rockets through Space*, the first serious treatment of astronautics in the United Kingdom. However, several established British scientific publications ridicule his book as the premature speculation of an unscientific imagination

✧ 1939–1945

Throughout World War II, nations use rockets and guided missiles of all sizes and shapes in combat. Of these, the most significant with respect to space exploration is the development of the liquid propellant V-2 rocket by the German army at Peenemünde under Wernher von Braun

✧ 1942

On October 3, the German A-4 rocket (later renamed Vengeance Weapon Two or V-2 Rocket) completes its first successful flight from the Peenemünde test site on the Baltic Sea. This is the birth date of the modern military ballistic missile

✧ 1944

In September, the German army begins a ballistic missile offensive by launching hundreds of unstoppable V-2 rockets (each carrying a one-ton high explosive warhead) against London and southern England

✧ 1945

Recognizing the war was lost, the German rocket scientist Wernher von Braun and key members of his staff surrender to American forces near Reutte, Germany in early May. Within months, U.S. intelligence teams, under Operation Paperclip, interrogate German rocket personnel and sort through carloads of captured documents and equipment. Many of these German scientists and engineers join von Braun in the United States to continue their rocket work. Hundreds of captured V-2 rockets are also disassembled and shipped back to the United States.

On May 5, the Soviet army captures the German rocket facility at Peenemünde and hauls away any remaining equipment and personnel. In the closing days of the war in Europe, captured German rocket technology and personnel helps set the stage for the great missile and space race of the cold war

On July 16, the United States explodes the world's first nuclear weapon. The test shot, code named Trinity, occurs in a remote portion of southern New Mexico and changes the face of warfare forever. As part of the cold-war confrontation between the United States and the former Soviet Union, the nuclear-armed ballistic missile will become the most powerful weapon ever developed by the human race.

In October, a then-obscure British engineer and writer, Arthur C. Clarke, suggests the use of satellites at geostationary orbit to support global communications. His article, in *Wireless World* "Extra-Terrestrial Relays," represents the birth of the communications satellite concept—an application of space technology that actively supports the information revolution

✧ 1946

On April 16, the U.S. Army launches the first American-adapted, captured German V-2 rocket from the White Sands Proving Ground in southern New Mexico.

Between July and August the Russian rocket engineer Sergei Korolev develops a stretched-out version of the German V-2 rocket. As part of his engineering improvements, Korolev increases the rocket engine's thrust and lengthens the vehicle's propellant tanks

✧ 1947

On October 30, Russian rocket engineers successfully launch a modified German V-2 rocket from a desert launch site near a place called Kapustin Yar. This rocket impacts about 199 miles (320 km) downrange from the launch site

✧ 1948

The September issue of the *Journal of the British Interplanetary Society* publishes the first in a series of four technical papers by L. R. Shepherd and A. V. Cleaver that explores the feasibility of applying nuclear energy to space travel, including the concepts of nuclear-electric propulsion and the nuclear rocket

✧ 1949

On August 29, the Soviet Union detonates its first nuclear weapon at a secret test site in the Kazakh Desert. Code-named First Lightning (Pervaya Molniya), the successful test breaks the nuclear-weapon monopoly enjoyed by the United States. It plunges the world into a massive nuclear arms race that includes the accelerated development of strategic ballistic missiles capable of traveling thousands of kilometers. Because they are well behind the United States in nuclear weapons technology, the leaders of the former Soviet Union decide to develop powerful, high-thrust rockets to carry their heavier, more primitive-design nuclear weapons. That decision gives the Soviet Union a major launch vehicle advantage when both superpowers decide to race into outer space (starting in 1957) as part of a global demonstration of national power

✧ 1950

On July 24, the United States successfully launches a modified German V-2 rocket with an American-designed WAC Corporal second-stage rocket from the U.S. Air Force's newly established Long Range Proving Ground at Cape Canaveral, Florida. The hybrid, multistage rocket (called Bumper 8) inaugurates the incredible sequence of military missile and space vehicle launches to take place from Cape Canaveral—the world's most famous launch site.

In November, British technical visionary Arthur C. Clarke publishes "Electromagnetic Launching as a Major Contribution to Space-Flight." Clarke's article suggests mining the Moon and launching the mined-lunar material into outer space with an electromagnetic catapult

✧ 1951

Cinema audiences are shocked by the science fiction movie *The Day the Earth Stood Still*. This classic story involves the arrival of a powerful, humanlike extraterrestrial and his robot companion, who come to warn the governments of the world about the foolish nature of their nuclear arms race. It is the first major science fiction story to portray powerful space aliens as friendly, intelligent creatures who come to help Earth.

Dutch-American astronomer Gerard Peter Kuiper suggests the existence of a large population of small, icy planetesimals beyond the orbit of Pluto—a collection of frozen celestial bodies now known as the Kuiper belt

✧ 1952

Collier's magazine helps stimulate a surge of American interest in space travel by publishing a beautifully illustrated series of technical articles written by space experts such as Wernher von Braun and Willey Ley. The first of the famous eight-part series appears on March 22 and is boldly titled "Man Will Conquer Space Soon." The magazine also hires the most influential space artist Chesley Bonestell to provide stunning color illustrations. Subsequent articles in the series introduce millions of American readers to the concept of a space station, a mission to the Moon, and an expedition to Mars

Wernher von Braun publishes *Das Marsprojekt* (The Mars project), the first serious technical study regarding a human-crewed expedition to

Mars. His visionary proposal involves a convoy of 10 spaceships with a total combined crew of 70 astronauts to explore the Red Planet for about one year and then return to Earth

✦ 1953

In August, the Soviet Union detonates its first thermonuclear weapon (a hydrogen bomb). This is a technological feat that intensifies the superpower nuclear arms race and increases emphasis on the emerging role of strategic, nuclear-armed ballistic missiles.

In October, the U.S. Air Force forms a special panel of experts, headed by John von Neumann to evaluate the American strategic ballistic missile program. In 1954, this panel recommends a major reorganization of the American ballistic missile effort

✦ 1954

Following the recommendations of John von Neumann, President Dwight D. Eisenhower gives strategic ballistic missile development the highest national priority. The cold war missile race explodes on the world stage as the fear of a strategic ballistic missile gap sweeps through the American government. Cape Canaveral becomes the famous proving ground for such important ballistic missiles as the Thor, Atlas, Titan, Minuteman, and Polaris. Once developed, many of these powerful military ballistic missiles also serve the United States as space launch vehicles. U.S. Air Force General Bernard Schriever oversees the time-critical development of the Atlas ballistic missile—an astonishing feat of engineering and technical management

✦ 1955

Walt Disney (the American entertainment visionary) promotes space travel by producing an inspiring three-part television series that includes appearances by noted space experts like Wernher von Braun. The first episode, "Man in Space," airs on March 9 and popularizes the dream of space travel for millions of American television viewers. This show, along with its companion episodes, "Man and the Moon" and "Mars and Beyond," make von Braun and the term *rocket scientist* household words

✦ 1957

On October 4, Russian rocket scientist Sergei Korolev with permission from Soviet premier Nikita S. Khrushchev uses a powerful military rocket to successfully place *Sputnik 1* (the world's first artificial satellite) into orbit around Earth. News of the Soviet success sends a political and technical shockwave across the United States. The launch of *Sputnik 1* marks the beginning of the Space Age. It also is the start of the great space race of the cold war—a period when people measure national strength and global prestige by accomplishments (or failures) in outer space.

On November 3, the Soviet Union launches *Sputnik 2*—the world's second artificial satellite. It is a massive spacecraft (for the time) that carries a live dog named Laika, which is euthanized at the end of the mission.

The highly publicized attempt by the United States to launch its first satellite with a newly designed civilian rocket ends in complete disaster on December 6. The Vanguard rocket explodes after rising only a few inches above its launch pad at Cape Canaveral. Soviet successes with *Sputnik 1* and *Sputnik 2* and the dramatic failure of the Vanguard rocket heighten American anxiety. The exploration and use of outer space becomes a highly visible instrument of cold-war politics

✧ 1958

On January 31, the United States successfully launches *Explorer 1*—the first American satellite in orbit around Earth. A hastily formed team from the U.S. Army Ballistic Missile Agency (ABMA) and Caltech's Jet Propulsion Laboratory (JPL), led by Wernher von Braun, accomplishes what amounts to a national prestige rescue mission. The team uses a military ballistic missile as the launch vehicle. With instruments supplied by Dr. James Van Allen of the State University of Iowa, *Explorer 1* discovers Earth's trapped radiation belts—now called the Van Allen radiation belts in his honor.

The National Aeronautics and Space Administration (NASA) becomes the official civilian space agency for the United States government on October 1. On October 7, the newly created NASA announces the start of the Mercury Project—a pioneering program to put the first American astronauts into orbit around Earth.

In mid-December, an entire Atlas rocket lifts off from Cape Canaveral and goes into orbit around Earth. The missile's payload compartment carries Project Score (Signal Communications Orbit Relay Experiment)—a prerecorded Christmas season message from President Dwight D. Eisenhower. This is the first time the human voice is broadcast back to Earth from outer space

✧ 1959

On January 2, the Soviet Union sends a 790 pound-mass (360-kg) spacecraft, *Lunik 1*, toward the Moon. Although it misses hitting the Moon by between 3,125 and 4,375 miles (5,000 and 7,000 km), it is the first human-made object to escape Earth's gravity and go in orbit around the Sun.

In mid-September, the Soviet Union launches *Lunik 2*. The 860 pound-mass (390-kg) spacecraft successfully impacts on the Moon and becomes the first human-made object to (crash-) land on another world. *Lunik 2* carries Soviet emblems and banners to the lunar surface.

On October 4, the Soviet Union sends *Lunik 3* on a mission around the Moon. The spacecraft successfully circumnavigates the Moon and takes the first images of the lunar farside. Because of the synchronous rotation of the Moon around Earth, only the near side of the lunar surface is visible to observers on Earth

✧ 1960

The United States launches the *Pioneer 5* spacecraft on March 11 into orbit around the Sun. The modest-sized (92 pound-mass [42-kg]) spherical American space probe reports conditions in interplanetary space between Earth and Venus over a distance of about 23 million miles (37 million km).

On May 24, the U.S. Air Force launches a MIDAS (Missile Defense Alarm System) satellite from Cape Canaveral. This event inaugurates an important American program of special military surveillance satellites intended to detect enemy missile launches by observing the characteristic infrared (heat) signature of a rocket's exhaust plume. Essentially unknown to the general public for decades because of the classified nature of their mission, the emerging family of missile surveillance satellites provides U.S. government authorities with a reliable early warning system concerning a surprise enemy (Soviet) ICBM attack. Surveillance satellites help support the national policy of strategic nuclear deterrence throughout the cold war and prevent an accidental nuclear conflict.

The U.S. Air Force successfully launches the *Discoverer 13* spacecraft from Vandenberg Air Force Base on August 10. This spacecraft is actually part of a highly classified Air Force and Central Intelligence Agency (CIA) reconnaissance satellite program called Corona. Started under special executive order from President Dwight D. Eisenhower, the joint agency spy satellite program begins to provide important photographic images of denied areas of the world from outer space. On August 18, *Discoverer 14* (also called *Corona XIV*) provides the U.S. intelligence community its first satellite-acquired images of the former Soviet Union. The era of satellite reconnaissance is born. Data collected by the spy satellites of the National Reconnaissance Office (NRO) contribute significantly to U.S. national security and help preserve global stability during many politically troubled times.

On August 12, NASA successfully launches the *Echo 1* experimental spacecraft. This large (100 foot [30.5 m] in diameter) inflatable, metalized balloon becomes the world's first passive communications satellite. At the dawn of space-based telecommunications, engineers bounce radio signals off the large inflated satellite between the United States and the United Kingdom.

The former Soviet Union launches *Sputnik 5* into orbit around Earth. This large spacecraft is actually a test vehicle for the new *Vostok* spacecraft that will soon carry cosmonauts into outer space. *Sputnik 5* carries two dogs, Strelka and Belka. When the spacecraft's recovery capsule functions properly the next day, these two dogs become the first living creatures to return to Earth successfully from an orbital flight

✧ 1961

On January 31, NASA launches a Redstone rocket with a Mercury Project space capsule on a suborbital flight from Cape Canaveral. The passenger astrochimp Ham is safely recovered down range in the Atlantic Ocean after reaching an altitude of 155 miles (250 km). This successful primate space mission is a key step in sending American astronauts safely into outer space.

The Soviet Union achieves a major space exploration milestone by successfully launching the first human being into orbit around Earth. Cosmonaut Yuri Gagarin travels into outer space in the *Vostok 1* spacecraft and becomes the first person to observe Earth directly from an orbiting space vehicle.

On May 5, NASA uses a Redstone rocket to send astronaut Alan B. Shepard, Jr., on his historic 15-minute suborbital flight into outer space from Cape Canaveral. Riding inside the Mercury Project *Freedom 7* space capsule, Shepard reaches an altitude of 115 miles (186 km) and becomes the first American to travel in space.

President John F. Kennedy addresses a joint session of the U.S. Congress on May 25. In an inspiring speech touching on many urgent national needs, the newly elected president creates a major space challenge for the United States when he declares: "I believe that this nation should commit itself to achieving the goal, before this decade is out, of landing a man on the Moon and returning him safely to Earth." Because of his visionary leadership, when American astronauts Neil A. Armstrong and Edwin E. "Buzz" Aldrin, Jr., step onto the lunar surface for the first time on July 20, 1969, the United States is recognized around the world as the undisputed winner of the cold-war space race

✧ 1962

On February 20, astronaut John Herschel Glenn, Jr., becomes the first American to orbit Earth in a spacecraft. An Atlas rocket launches the NASA Mercury Project *Friendship 7* space capsule from Cape Canaveral. After completing three orbits, Glenn's capsule safely splashes down in the Atlantic Ocean.

In late August, NASA sends the *Mariner 2* spacecraft to Venus from Cape Canaveral. *Mariner 2* passes within 21,700 miles (35,000 km) of the planet on December 14, 1962—thereby becoming the world's first successful interplanetary space probe. The spacecraft observes very high surface temperatures (~800°F [430°C]). These data shatter pre–space age visions about Venus being a lush, tropical planetary twin of Earth.

During October, the placement of nuclear-armed Soviet offensive ballistic missiles in Fidel Castro's Cuba precipitates the Cuban Missile Crisis. This dangerous superpower confrontation brings the world perilously close to nuclear warfare. Fortunately, the crisis dissolves when Premier Nikita S. Khrushchev withdraws the Soviet ballistic missiles after much skillful political maneuvering by President John F. Kennedy and his national security advisers

✧ 1964

On November 28, NASA's *Mariner 4* spacecraft departs Cape Canaveral on its historic journey as the first spacecraft from Earth to visit Mars. It successfully encounters the Red Planet on July 14, 1965 at a flyby distance of about 6,100 miles (9,800 km). *Mariner 4*'s closeup images reveal a barren, desertlike world and quickly dispel any pre–space age notions about the existence of ancient Martian cities or a giant network of artificial canals

✧ 1965

A Titan II rocket carries astronauts Virgil "Gus" I. Grissom and John W. Young into orbit on March 23 from Cape Canaveral, inside a two-person Gemini Project spacecraft. NASA's *Gemini 3* flight is the first crewed mission for the new spacecraft and marks the beginning of more sophisticated space activities by American crews in preparation for the Apollo Project lunar missions

✧ 1966

The former Soviet Union sends the *Luna 9* spacecraft to the Moon on January 31. The 220 pound-mass (100-kg) spherical spacecraft soft lands in the Ocean of Storms region on February 3, rolls to a stop, opens four petal-like covers, and then transmits the first panoramic television images from the Moon's surface.

The former Soviet Union launches the *Luna 10* to the Moon on March 31. This massive (3,300 pound-mass [1,500-kg]) spacecraft becomes the first human-made object to achieve orbit around the Moon.

On May 30, NASA sends the *Surveyor 1* lander spacecraft to the Moon. The versatile robot spacecraft successfully makes a soft landing (June 1) in the Ocean of Storms. It then transmits over 10,000 images from the lunar surface and performs numerous soil mechanics experiments in preparation for the Apollo Project human landing missions.

In mid-August, NASA sends the *Lunar Orbiter 1* spacecraft to the Moon from Cape Canaveral. It is the first of five successful missions to collect detailed images of the Moon from lunar orbit. At the end of each mapping mission the orbiter spacecraft is intentionally crashed into the Moon to prevent interference with future orbital activities

✧ 1967

On January 27, disaster strikes NASA's Apollo Project. While inside their *Apollo 1* spacecraft during a training exercise on Launch Pad 34 at Cape Canaveral, astronauts Virgil "Gus" I. Grissom, Edward H. White, Jr., and Roger B. Chaffee are killed when a flash fire sweeps through their space-craft. The Moon landing program was delayed by 18 months, while major design and safety changes are made in the Apollo Project spacecraft.

On April 23, tragedy also strikes the Russian space program when the Soviets launch cosmonaut Vladimir Komarov in the new *Soyuz* (union) spacecraft. Following an orbital mission plagued with difficulties, Koma-rov dies (on April 24) during reentry operations, when the spacecraft's parachute fails to deploy properly and the vehicle hits the ground at high speed

✧ 1968

On December 21, NASA's *Apollo 8* spacecraft (command and service modules only) departs Launch Complex 39 at the Kennedy Space Center during the first flight of mighty Saturn V launch vehicle with a human crew as part of the payload. Astronauts Frank Borman, James Arthur Lovell, Jr., and William A. Anders become the first people to leave Earth's gravitational influence. They go into orbit around the Moon and capture images of an incredibly beautiful Earth "rising" above the starkly barren lunar horizon—pictures that inspire millions and stimulate an emerging environmental movement. After 10 orbits around the Moon, the first lunar astronauts return safely to Earth on December 27

✧ 1969

The entire world watches as NASA's *Apollo 11* mission leaves for the Moon on July 16 from the Kennedy Space Center. Astronauts Neil A. Armstrong, Michael Collins, and Edwin E. "Buzz" Aldrin, Jr., make a long-held dream of humanity a reality. On July 20, American astronaut Neil Armstrong cau-tiously descends the steps of the lunar excursion module's ladder and steps on the lunar surface, stating, "One small step for a man, one giant leap for mankind!" He and Buzz Aldrin become the first two people to walk on another world. Many people regard the Apollo Project lunar landings as the greatest technical accomplishment in all of human history

✧ 1970

NASA's *Apollo 13* mission leaves for the Moon on April 11. Suddenly, on April 13, a life-threatening explosion occurs in the service module portion of the Apollo spacecraft. Astronauts James A. Lovell, Jr., John Leonard Swigert, and Fred Wallace Haise, Jr., must use their lunar excursion module (LEM) as a lifeboat. While an anxious world waits and listens, the crew skillfully maneuvers their disabled spacecraft around the Moon. With critical supplies running low, they limp back to Earth on a free-return trajectory. At just the right moment on April 17, they abandon the LEM *Aquarius* and board the Apollo Project spacecraft (command module) for a successful atmospheric reentry and recovery in the Pacific Ocean

✧ 1971

On April 19, the former Soviet Union launches the first space station (called *Salyut 1*). It remains initially uncrewed because the three-cosmonaut crew of the *Soyuz 10* mission (launched on April 22) attempts to dock with the station but cannot go on board

✧ 1972

In early January, President Richard M. Nixon approves NASA's space shuttle program. This decision shapes the major portion of NASA's program for the next three decades.

On March 2, an Atlas-Centaur launch vehicle successfully sends NASA's *Pioneer 10* spacecraft from Cape Canaveral on its historic mission. This far-traveling robot spacecraft becomes the first to transit the main-belt asteroids, the first to encounter Jupiter (December 3, 1973) and by crossing the orbit of Neptune on June 13, 1983 (which at the time was the farthest planet from the Sun) the first human-made object ever to leave the planetary boundaries of the solar system. On an interstellar trajectory, *Pioneer 10* (and its twin, *Pioneer 11*) carries a special plaque, greeting any intelligent alien civilization that might find it drifting through interstellar space millions of years from now.

On December 7, NASA's *Apollo 17* mission, the last expedition to the Moon in the 20th century, departs from the Kennedy Space Center, propelled by a mighty Saturn V rocket. While astronaut Ronald E. Evans remains in lunar orbit, fellow astronauts Eugene A. Cernan and Harrison H. Schmitt become the 11th and 12th members of the exclusive Moon walkers club. Using a lunar rover, they explore the Taurus-Littrow region. Their safe return to Earth on December 19 brings to a close one of the epic periods of human exploration

✧ 1973

In early April, while propelled by Atlas-Centaur rocket, NASA's *Pioneer 11* spacecraft departs on an interplanetary journey from Cape Canaveral. The spacecraft encounters Jupiter (December 2, 1974) and then uses a gravity assist maneuver to establish a flyby trajectory to Saturn. It is the first spacecraft to view Saturn at close range (closest encounter on September 1, 1979) and then follows a path into interstellar space.

On May 14, NASA launches *Skylab*—the first American space station. A giant Saturn V rocket is used to place the entire large facility into orbit in a single launch. The first crew of three American astronauts arrives on May 25 and makes the emergency repairs necessary to save the station, which suffered damage during the launch ascent. Astronauts Charles (Pete) Conrad, Jr., Paul J. Weitz, and Joseph P. Kerwin stay onboard for 28 days. They are replaced by astronauts Alan L. Bean, Jack R. Lousma, and Owen K. Garriott, who arrive on July 28 and live in space for about 59 days. The final *Skylab* crew (astronauts Gerald P. Carr, William R. Pogue, and Edward G. Gibson) arrive on November 11 and resided in the station until February 8, 1974—setting a space endurance record (for the time) of 84 days. NASA then abandons *Skylab*.

In early November, NASA launches the *Mariner 10* spacecraft from Cape Canaveral. It encounters Venus (February 5, 1974) and uses a gravity assist maneuver to become the first spacecraft to investigate Mercury at close range

✧ 1975

In late August and early September, NASA launches the twin *Viking 1* (August 20) and *Viking 2* (September 9) orbiter/lander combination spacecraft to the Red Planet from Cape Canaveral. Arriving at Mars in 1976, all Viking Project spacecraft (two landers and two orbiters) perform exceptionally well—but the detailed search for microscopic alien life-forms on Mars remains inconclusive

✧ 1977

On August 20, NASA sends the *Voyager 2* spacecraft from Cape Canaveral on an epic grand tour mission during which it encounters all four giant planets and then departs the solar system on an interstellar trajectory. Using the gravity assist maneuver, *Voyager 2* visits Jupiter (July 9, 1979), Saturn (August 25, 1981), Uranus (January 24, 1986), and Neptune (August 25, 1989). The resilient, far-traveling robot spacecraft (and its twin *Voyager 1*) also carries a special interstellar message from Earth—a digital record entitled *The Sounds of Earth.*

On September 5, NASA sends the *Voyager 1* spacecraft from Cape Canaveral on its fast trajectory journey to Jupiter (March 5, 1979), Saturn (March 12, 1980), and beyond the solar system

✦ 1978
In May, the British Interplanetary Society releases its Project Daedalus report—a conceptual study about a one-way robot spacecraft mission to Barnard's star at the end of the 21st century

✦ 1979
On December 24, the European Space Agency successfully launches the first Ariane 1 rocket from the Guiana Space Center in Kourou, French Guiana

✦ 1980
India's Space Research Organization successfully places a modest 77 pound-mass (35 kg) test satellite (called *Rohini*) into low Earth orbit on July 1. The launch vehicle is a four-stage, solid propellant rocket manufactured in India. The SLV-3 (Standard Launch Vehicle-3) gives India independent national access to outer space

✦ 1981
On April 12, NASA launches the space shuttle *Columbia* on its maiden orbital flight from Complex 39-A at the Kennedy Space Center. Astronauts John W. Young and Robert L. Crippen thoroughly test the new aerospace vehicle. Upon reentry, it becomes the first spacecraft to return to Earth by gliding through the atmosphere and landing like an airplane. Unlike all previous onetime use space vehicles, *Columbia* is prepared for another mission in outer space

✦ 1986
On January 24, NASA's *Voyager 2* spacecraft encounters Uranus.

On January 28, the space shuttle *Challenger* lifts off from the NASA Kennedy Space Center on its final voyage. At just under 74 seconds into the STS 51-L mission, a deadly explosion occurs, killing the crew and destroying the vehicle. Led by President Ronald Reagan, the United States mourns seven astronauts lost in the *Challenger* accident

✦ 1988
On September 19, the State of Israel uses a Shavit (comet) three-stage rocket to place the country's first satellite (called *Ofeq 1*) into an unusual east-to-west orbit—one that is opposite to the direction of Earth's rotation but necessary because of launch safety restrictions.

As the *Discovery* successfully lifts off on September 29 for the STS-26 mission, NASA returns the space shuttle to service following a 32-month hiatus after the *Challenger* accident

✧ 1989

On August 25, the *Voyager 2* spacecraft encounters Neptune

✧ 1994

In late January, a joint Department of Defense and NASA advanced technology demonstration spacecraft, *Clementine*, lifts off for the Moon from Vandenberg Air Force Base. Some of the spacecraft's data suggest that the Moon may actually possess significant quantities of water ice in its permanently shadowed polar regions

✧ 1995

In February, during NASA's STS-63 mission, the space shuttle *Discovery* approaches (encounters) the Russian *Mir* space station as a prelude to the development of the *International Space Station*. Astronaut Eileen Marie Collins serves as the first female shuttle pilot.

On March 14, the Russians launch the *Soyuz TM-21* spacecraft to the *Mir* space station from the Baikanour Cosmodrome. The crew of three includes American astronaut Norman Thagard—the first American to travel into outer space on a Russian rocket and the first to stay on the *Mir* space station. The *Soyuz TM-21* cosmonauts also relieve the previous *Mir* crew, including cosmonaut Valeri Polyakov, who returns to Earth on March 22 after setting a world record for remaining in space for 438 days.

In late June, NASA's space shuttle *Atlantis* docks with the Russian *Mir* space station for the first time. During this shuttle mission (STS-71), *Atlantis* delivers the *Mir 19* crew (cosmonauts Anatoly Solovyev and Nikolai Budarin) to the Russian space station and then returns the *Mir 18* crew back to Earth—including American astronaut Norman Thagard, who has just spent 115 days in space onboard the *Mir*. The Shuttle-*Mir* docking program is the first phase of the *International Space Station*. A total of nine shuttle-*Mir* docking missions will occur between 1995 and 1998

✧ 1998

In early January, NASA sends the *Lunar Prospector* to the Moon from Cape Canaveral. Data from this orbiter spacecraft reinforces previous hints that the Moon's polar regions may contain large reserves of water ice in a mixture of frozen dust lying at the frigid bottom of some permanently shadowed craters.

In early December, the space shuttle *Endeavour* ascends from the NASA Kennedy Space Center on the first assembly mission of the *International Space Station*. During the STS-88 shuttle mission, *Endeavour* performs a rendezvous with the previously launched Russian-built *Zarya* (sunrise) module. An international crew connects this module with the American-built *Unity* module carried in the shuttle's cargo bay

✧ 1999

In July, astronaut Eileen Marie Collins serves as the first female space shuttle commander (STS-93 mission) as the *Columbia* carries NASA's *Chandra X-ray Observatory* into orbit

✧ 2001

NASA launches the *Mars Odyssey 2001* mission to the Red Planet in early April—the spacecraft successfully orbits the planet in October

✧ 2002

On May 4, NASA successfully launches its *Aqua* satellite from Vandenberg Air Force Base. This sophisticated Earth-observing spacecraft joins the *Terra* spacecraft in performing Earth system science studies.

On October 1, the United States Department of Defense forms the U.S. Strategic Command (USSTRATCOM) as the control center for all American strategic (nuclear) forces. USSTRATCOM also conducts military space operations, strategic warning and intelligence assessment, and global strategic planning

✧ 2003

On February 1, while gliding back to Earth after a successful 16-day scientific research mission (STS-107), the space shuttle *Columbia* experiences a catastrophic reentry accident at an altitude of about 63 km over the Western United States. Traveling at 18 times the speed of sound, the orbiter vehicle disintegrates, taking the lives of all seven crew members: six American astronauts (Rick Husband, William McCool, Michael Anderson, Kalpana Chawla, Laurel Clark, and David Brown) and the first Israeli astronaut (Ilan Ramon).

NASA's Mars Exploration Rover (MER) *Spirit* is launched by a Delta II rocket to the Red Planet on June 10. *Spirit,* also known as MER-A, arrives safely on Mars on January 3, 2004 and begins its teleoperated surface exploration mission under the supervision of mission controllers at the NASA Jet Propulsion Laboratory.

NASA launches the second Mars Exploration Rover, called *Opportunity,* using a Delta II rocket launch, which lifts off from Cape Canaveral Air Force Station on July 7, 2003. *Opportunity,* also called MER-B, successfully lands on Mars on January 24, 2004, and starts its teleoperated surface exploration mission under the supervision of mission controllers at the NASA Jet Propulsion Laboratory

✧ 2004

On July 1, NASA's *Cassini* spacecraft arrives at Saturn and begins its four-year mission of detailed scientific investigation.

In mid-October, the Expedition 10 crew, riding a Russian launch vehicle from Baikonur Cosmodrome, arrives at the *International Space Station* and the Expedition 9 crew returns safely to Earth.

On December 24, the 703 pound-mass (319-kg) *Huygens* probe successfully separates from the *Cassini* spacecraft and begins its journey to Saturn's moon, Titan

✧ 2005

On January 14, the *Huygens* probe enters the atmosphere of Titan and successfully reaches the surface some 147 minutes later. *Huygens* is the first spacecraft to land on a moon in the outer solar system.

On July 4, NASA's Deep Impact mission successfully encountered Comet Tempel 1.

NASA successfully launched the space shuttle *Discovery* on the STS-114 mission on July 26 from the Kennedy Space Center in Florida. After docking with the *International Space Station*, the *Discovery* returned to Earth and landed at Edwards AFB, California, on August 9.

On August 12, NASA launched the *Mars Reconnaissance Orbiter* from Cape Canaveral AFS, Florida.

On September 19, NASA announced plans for a new spacecraft designed to carry four astronauts to the Moon and to deliver crews and supplies to the *International Space Station*. NASA also introduced two new, shuttle-derived launch vehicles: a crew-carrying rocket and a cargo-carrying, heavy-lift rocket.

The Expedition 12 crew (Commander William McArthur and Flight Engineer Valery Tokarev) arrived at the *International Space Station* on October 3 and replaced the Expedition 11 crew.

The People's Republic of China successfully launched its second human spaceflight mission, called *Shenzhou 6*, on October 12. Two taikonauts, Fei Junlong and Nie Haisheng, traveled in space for almost five days and made 76 orbits of Earth before returning safely to Earth, making a soft, parachute-assisted landing in northern Inner Mongolia

✧ 2006

On January 15, the sample package from NASA's *Stardust* spacecraft, containing comet samples, successfully returned to Earth.

NASA launched the *New Horizons* spacecraft from Cape Canaveral on January 19 and successfully sent this robot probe on its long one-way mission to conduct a scientific encounter with the Pluto system (in 2015) and then to explore portions of the Kuiper belt that lie beyond.

Follow-up observations by NASA's *Hubble Space Telescope*, reported on February 22, have confirmed the presence of two new moons around the distant planet Pluto. The moons, tentatively called S/2005 P 1 and S/2005 P 2, were first discovered by *Hubble* in May 2005, but the science team wanted to further examine the Pluto system to characterize the orbits of the new moons and validate the discovery.

NASA scientists announced on March 9 that the *Cassini* spacecraft may have found evidence of liquid water reservoirs that erupt in Yellowstone Park–like geysers on Saturn's moon Enceladus.

On March 10, NASA's *Mars Reconnaissance Orbiter* successfully arrived at Mars and began a six-month-long process of adjusting and trimming the shape of its orbit around the Red Planet prior to performing its operational mapping mission.

The Expedition 13 crew (Commander Pavel Vinogradov and Flight Engineer Jeff Williams) arrived at the *International Space Station* on April 1 and replaced the Expedition 12 crew. Joining them for several days before returning back to Earth with the Expedition 12 crew was Brazil's first astronaut, Marcos Pontes.

On August 24, members of the International Astronomical Union (IAU) met for the organization's 2006 General Assembly in Prague, Czech Republic. After much debate, the 2,500 assembled professional astronomers decided (by vote) to demote Pluto from its traditional status as one of the nine major planets and place the object into a new class, called a dwarf planet. The IAU decision now leaves the solar system with eight major planets and three dwarf planets: Pluto (which serves as the prototype dwarf planet), Ceres (the largest asteroid), and the large, distant Kuiper belt object identified as 2003 UB313 (nicknamed "Xena"). Astronomers anticipate the discovery of other dwarf planets in the distant parts of the solar system.

On September 9, NASA successfully launched the space shuttle *Atlantis* on the STS-115 mission. The shuttle crew delivered and installed the P3/P4 truss structure on the *International Space Station*. Following its orbital mission, the *Atlantis* landed on September 21 at the Kennedy Space Center.

The Expedition 14 crew (Commander Michael Lopez-Alegria and Flight Engineer Mikhail Tyurin) arrived at the *International Space Station* on September 20 and the Expedition 13 crew returned safely to Earth eight days later.

NASA used a spectacular nighttime launch to send the space shuttle *Discovery* into orbit on December 9. During the STS-116 mission, the crew of *Discovery* docked with the *International Space Station* to deliver and install the P5 truss structure.

Glossary

abiotic   Abiotic does not involve living things and is not produced by living organisms.

abundance of elements (in the universe)   Stellar spectra provide an estimate of the cosmic abundance of elements as a percent of the total mass of the universe. The 10 most common elements are: hydrogen (H) 73.5 percent of the total mass, helium (He) 24.9 percent, oxygen (O) 0.7 percent, carbon (C) 0.3 percent, iron (Fe) 0.15 percent, neon (Ne) 0.12 percent, nitrogen (N) 0.10 percent, silicon (Si) 0.07 percent, magnesium (Mg) 0.05 percent, and sulfur (S) 0.04 percent.

acceleration of gravity   The local acceleration is due to gravity on or near the surface of a planet. On Earth, the acceleration due to gravity (symbol: g) of a free falling object has the standard value of 32.1740 feet per second per second (9.80665 m/s^2) by international agreement.

acronym   An acronym is word that is formed from the first letters of a name, such as *HST,* which means the *Hubble Space Telescope;* or a word that is formed by combining the initial parts of a series of words, such as *lidar,* which means *li*ght *d*etection *a*nd *r*anging. Acronyms are frequently used in the space technology and astronomy.

acute radiation syndrome (ARS)   This acute organic disorder follows exposure to relatively severe doses of ionizing radiation. A person will initially experience nausea, diarrhea, or blood cell changes. In the later stages, loss of hair, hemorrhaging, and possibly death can take place. Radiation-dose equivalent values of about 450 to 500 rem (4.5 to 5 sievert) will prove fatal to 50 percent of the exposed individuals in a large general population. This is also called radiation sickness.

aerospace vehicle   An aerospace vehicle is capable of operating both within Earth's sensible (measurable) atmosphere and in outer space. The space-shuttle orbiter vehicle is an example.

age of the Earth   Planet Earth is very old, about 4.5 billion years or more, according to recent estimates. Most of the evidence for an ancient Earth is contained in the rocks that form the planet's crust. The ages of Earth and Moon rocks and of meteorites are measured by the decay of long-lived radioactive isotopes of elements that occur naturally in rocks and minerals and that decay with half-lives of 700 million to more than 100 billion years into stable isotopes of other elements. In Western Australia, single zircon crystals found in younger sedimentary rocks have radiometric ages of as much as 4.3 billion years, making these tiny crystals the oldest materials to be found on Earth so far. The oldest dated Moon rocks, however, have ages between 4.4 and 4.5 billion years and provide a minimum age for the formation of Earth's nearest planetary neighbor.

air   Air is the overall mixture of gases that make up Earth's atmosphere, primarily nitrogen (N_2) at 78 percent (by volume), oxygen (O_2) at 21 percent, argon (Ar) at 0.9 percent, and carbon dioxide (CO_2) at 0.03 percent. Sometimes, aerospace engineers use this word for the breathable gaseous mixture found inside the crew compartment of a space vehicle or in the pressurized habitable environment of a space station.

alien life-form (ALF)   An alien life-form general, though at present hypothetical, expression for extraterrestrial life, especially life that exhibits some degree of intelligence.

Alpha Centauri   The closest star system, about 4.3 light-years away, it is actually a triple-star system, with two stars orbiting around each other and a third star, called Proxima Centauri, revolving around the pair at some distance.

alphanumeric (*alpha*bet plus *numeric*)   Alphanumeric includes letters and numerical digits, as for example, the term JEN75WX11.

amino acid   This acid contains the amino (NH_2) group, which is a group of molecules that is necessary for life. More than 80 amino acids are presently known, but only some 20 occur naturally in living organisms, where they serve as the building blocks of proteins. On Earth, many microorganisms and plants can synthesize amino acids from simple inorganic compounds.

ancient astronaut theory   The ancient astronaut theory is an (unproven) hypothesis that Earth was visited in the past by a race of intelligent, extra-terrestrial beings who were exploring this portion of the Milky Way galaxy.

android   Android is a term from science fiction that describes an intel-ligent robot with near-human form or features.

Andromeda galaxy   The Great Spiral galaxy (or M31) in the constel-lation of Andromeda, about 2.2 million light-years way, the Andromeda galaxy is the most distant object that is visible to the naked eye and is the closest spiral galaxy to the Milky Way galaxy.

antenna   An antenna is a device that is used to detect, collect, or transmit radio waves. A radio telescope is a large receiving antenna, while many spacecraft have both a directional antenna and an omnidirectional antenna to transmit (downlink) telemetry and to receive (uplink) instructions.

anthropic principle   The anthropic principle is a controversial hypoth-esis in modern cosmology that suggests that the universe evolved in just the right way after the big bang event to allow for the emergence of human life.

apastron   Apastron is the point in a body's orbit around a star at which the body is at a maximum distance from the star. *Compare with* PERIASTRON.

Apollo Project   The U.S. effort in the 1960s and early 1970s to place astronauts successfully on the surface of the Moon and return them safely to Earth, the project was initiated in May 1961 by President John F. Ken-nedy (1917–63) in response to a growing space-technology challenge from the former Soviet Union. Managed by NASA, the *Apollo 8* mission sent the first three humans to the vicinity of the Moon in December 1968. The *Apollo 11* mission involved the first human landing on another world (July 20, 1969). *Apollo 17,* the last lunar landing mission under this project, took place in December 1972. The project is often considered to be one of the greatest technical accomplishments in all human history.

apolune   An apolune is that point in an orbit around the Moon of a spacecraft launched from the lunar surface that is farthest from the Moon. *Compare with* PERILUNE.

Arecibo Interstellar Message   To help inaugurate the powerful radio/radar telescope of the Arecibo Observatory, an interstellar radio-wave message of friendship was beamed to the fringes of the Milky Way galaxy on November 16, 1974. Scientists sent a special radio signal toward the Great Cluster in Hercules—a globular cluster that lies about 25,000 light-years away from Earth and contains about 300,000 stars within a radius of approximately 18 light-years.

Arecibo Observatory   The world's largest radio/radar telescope with a 1,000-foot (305-m) diameter dish, the Arecibo Observatory is located in a large, bowl-shaped natural depression in the tropical jungles of Puerto Rico.

artificial gravity   This simulated gravity condition is established within a spacecraft, space station, or large space settlement. Rotating the human-occupied space system about an axis creates an artificial condition, since the centrifugal force generated by the rotation produces effects similar to the force of gravity within the vehicle.

artificial intelligence (AI)   Information-processing (including thinking and perceiving), artificial intelligence is performed by machines in a way that imitates (to some extent) the mental activities that are performed by the human brain. Advances in AI will allow "very smart" robot spacecraft to explore distant alien worlds with minimal human supervision.

asteroid   Small, solid rocky objects that orbit the Sun but are independent of any major planet, most asteroids (or minor planets) are found in the main asteroid belt between the orbits of Mars and Jupiter. The largest asteroid is Ceres, which was discovered in 1801 by the Italian astronomer Giuseppe Piazzi (1746–1826). Earth-crossing asteroids or near-Earth asteroids (NEAs) have orbits that take them near or across Earth's orbit around the Sun and are divided into the Aten, Apollo, and Amor groups.

astro-   The prefix *astro-* means star or (by extension) outer space or celestial; for example, *astronaut, astronautics,* or *astrophysics.*

astrobiology   This search for and study of living organisms found on celestial bodies beyond Earth is also called exobiology.

astronaut   Within the U.S. space program, an astronaut is a person who travels in outer space or a person who flies in an aerospace vehicle to an altitude of more than 50 miles (80 km). The word comes from a combination of two ancient Greek words that literally mean "star" (*astro*) "sailor or traveler" (*naut*). *Compare with* COSMONAUT.

astronomical unit (AU)   A convenient unit of distance that is defined as the semimajor axis of Earth's orbit around the Sun, 1 astronomical unit (AU), the average distance between Earth and the Sun, is equal to approximately 92.2×10^6 miles (149.6×10^6 km) or 499.01 light-seconds.

atmosphere   Atmosphere takes two forms: 1. (cabin) The breathable environment inside a human-occupied space capsule, aerospace vehicle, spacecraft, or space station. 2. (planetary) The gravitationally bound gaseous envelope that forms an outer region around a planet or other celestial body.

atmospheric probe   An atmospheric probe is the special collection of scientific instruments (usually released by a mother spacecraft) for determining the pressure, composition, and temperature of a planet's atmosphere at different altitudes. An example is the probe released by NASA's *Galileo* spacecraft in December 1995. As it plunged into Jupiter's atmosphere, the probe successfully transmitted its scientific data to the *Galileo* spacecraft (the mother spacecraft) for about 58 minutes.

Barnard's star   A red dwarf star approximately six light-years from the Sun, Barnard's star the fourth-nearest star to the solar system. Discovered in 1916 by the U.S. astronomer Edward Emerson Barnard (1857–1923), it has the largest proper motion (some 10.3 seconds of arc per year) of any known star.

Bernal sphere   This large, spherically shaped space settlement was first proposed in 1929 by the Irish physicist and writer John Desmond Bernal (1910–71).

big bang (theory)   A contemporary theory in cosmology, the big bang theory concerns the origin of the universe. It suggests that a very large, ancient explosion started space and time of the present universe, which has been expanding ever since.

big crunch   Within the closed universe model of cosmology, the big crunch is the postulated end state that occurs after the present universe expands to its maximum physical dimensions and then collapses in on itself under the influence of gravitation, eventually reaching an infinitely dense end point, or singularity.

binary star system   A pair of stars that orbit around a common center of mass and are bound together by their mutual gravitation is a binary star system.

biogenic elements   Biogenic elements generally are considered by scientists (astrobiologists) as essential for all living systems and include hydrogen (H), carbon (C), nitrogen (N), oxygen (O), sulfur (S), and phosphorous (P). The availability of the chemical compound, water (H_2O), is also considered necessary for life both here on Earth and possibly elsewhere in the universe.

biosphere   The life zone of a planetary body—for example, that part of the Earth system inhabited by living organisms—the biosphere on this planet includes portions of the atmosphere, the hydrosphere, the cryosphere, and surface regions of the solid Earth.

biotelemetry   The remote measurement of life functions, biotelemetry is data from biosensors that are attached to an astronaut or cosmonaut and that are sent back to Earth (as telemetry) for the purposes of spacecrew health monitoring and evaluation by medical experts and mission managers.

black dwarf   A black dwarf consists of the cold remains of a white dwarf star that no longer emits visible radiation or a nonradiating ball of interstellar gas that has contracted under gravitation but contains too little mass to initiate nuclear fusion.

black hole   An incredibly compact, gravitationally collapsed mass from which nothing (light, matter, or any other kind of information) can escape. Astrophysicists believe that a black hole is the natural end product when a very massive star dies and collapses beyond a certain critical dimension.

blueshift   *See* DOPPLER SHIFT.

brown dwarf   A brown dwarf is a very low-luminosity, substellar (almost a star) celestial object that contains starlike material (that is, hydrogen and helium) but has too low a mass (typically 1 to 10 percent of a solar mass) to allow its core to initiate thermonuclear fusion (hydrogen burning).

Callisto   *See* GALILEAN SATELLITES.

canali The Italian word *canali* (channels) was used in 1877 by the Italian astronomer Giovanni Virginio Schiaparelli (1835–1910) to describe natural surface features that he observed on Mars. Subsequent pre–space age investigators, including the U.S. astronomer Percival Lowell (1855–1916), took the Italian word quite literally as meaning *canals* and sought additional evidence of an intelligent civilization on Mars. Since the 1960s, many spacecraft have visited Mars, dispelling such popular speculations and revealing no evidence of any Martian canals that had been constructed by intelligent beings.

Cape Canaveral The region on Florida's east central coast from which the U.S. Air Force and NASA have launched more than 3,000 rockets since 1950, Cape Canaveral Air Force Station (CCAFS) is the major East Coast launch site for the Department of Defense, while the adjacent NASA Kennedy Space Center is the spaceport for the fleet of space-shuttle vehicles.

Cassini mission The joint NASA-European Space Agency planetary exploration Cassini mission to Saturn was launched from Cape Canaveral on October 15, 1997. Since July 2004, the *Cassini* spacecraft has performed detailed studies of Saturn, its rings, and its moons. The *Cassini* mother spacecraft also delivered the *Huygens* probe, which successfully plunged into the nitrogen-rich atmosphere of Titan (Saturn's largest moon) on January 14, 2005. The mother spacecraft is named after the Italian-French astronomer Giovanni Cassini (1625–1712); the Titan probe after the Dutch astronomer Christiaan Huygens (1629–95).

celestial body A heavenly body, a celestial body is any aggregation of matter in outer space that constitutes a unit for study in astronomy, such as planets, moons, comets, asteroids, stars, nebulas, and galaxies.

Chryse Planitia A large plain on Mars that is characterized by many ancient channels that could have once contained flowing surface water, Chryse Planitia was the landing site for NASA's *Viking 1* lander (robot spacecraft) in July 1976.

cislunar Of or pertaining to phenomena, projects, or activities that happen in the region of outer space between Earth and the Moon, *cislunar* comes from the Latin word *cis,* meaning "on this side," and *lunar,* which means "of or pertaining to the Moon." Therefore, it means "on this side of the Moon."

clean room   A clean room is a controlled work environment for spacecraft and aerospace systems in which dust, temperature, and humidity are carefully controlled during the fabrication, assembly, and/or testing of critical components.

closed ecological life-support system (CELSS)   This system can provide for the maintenance of life in an isolated living chamber or facility through complete reuse of the materials that are available within the chamber or facility.

close encounter (CE)   A close encounter is a postulated interaction with an unidentified flying object (UFO). *See also* UNIDENTIFIED FLYING OBJECT.

cold war   The ideological conflict between the United States and the former Soviet Union from approximately 1946 to 1989, the cold war involved rivalry, mistrust, and hostility just short of overt military action. The tearing down of the Berlin Wall in November 1989 generally is considered as the (symbolic) end of the cold-war period.

coma   The gaseous envelope that surrounds the nucleus of a comet is called a coma.

comet   A comet is a dirty ice "rock" consisting of dust, frozen water, and gases that orbits the Sun. As a comet approaches the inner solar system from deep space, solar radiation causes its frozen materials to vaporize (sublime), creating a coma and a long tail of dust and ions. Scientists think that these icy planetesimals are the remainders of the primordial material from which the outer planets were formed billions of years ago. *See also* KUIPER BELT and OORT CLOUD.

comet Halley   (1P/Halley) The most famous periodic comet, the comet Halley is named after the British astronomer Edmond Halley (1656–1742), who successfully predicted its 1758 return. Reported since 240 B.C.E., this comet reaches perihelion approximately every 76 years. During its most-recent inner solar-system appearance, an international fleet of five different robot spacecraft, including the *Giotto* spacecraft, performed scientific investigations that supported the dirty ice rock model of a comet's nucleus.

Condon, Edward Uhler   (1902–1974) The U.S. theoretical physicist Edward Uhler Condon served as the director of an investigation that was sponsored by the U.S. Air Force (USAF) concerning unidentified flying object (UFO) sighting reports. These reports were accumulated between

1948 and 1966 under USAF Project Blue Book (and its predecessors). *See also* UNIDENTIFIED FLYING OBJECT.

continuously habitable zone (CHZ)   The region around a star in which one or several planets (or possibly their moons) can maintain conditions appropriate for the emergence and sustained existence of life is the continuously habitable zone. One important characteristic of a planet in the CHZ is that its environmental conditions support the retention of significant amounts of liquid water on the planetary surface.

coorbital   Sharing the same or very similar orbit; for example, during a rendezvous operation the chaser spacecraft and its cooperative target are said to be coorbital.

cosmic rays   Extremely energetic particles, or cosmic rays (usually bare atomic nuclei), move through outer space at speeds just below the speed of light and bombard Earth from all directions.

cosmological principle   The cosmological principle, or hypothesis, states that the expanding universe is isotropic and homogeneous. In other words, there is no special location for observing the universe, and all observers anywhere in the universe see the same recession of distant galaxies.

cosmology   The study of the origin, the evolution, and the structure of the universe, contemporary cosmology centers around the big bang hypothesis—a theory stating that about 15 billion (10^9) years ago, the universe began in a great explosion and has been expanding ever since.

cosmonaut   The title *cosmonaut* was given by Russia (formerly the Soviet Union) to its space travelers or *astronauts*.

crater   A bowl-shaped topographic depression with steep slopes on the surface of a planet or moon, a crater will be one of two general types: impact crater (as formed by an asteroid, comet, or meteoroid strike) and eruptive crater (as formed when a volcano erupts).

cyborg   *Cyborg* is a contraction of the expression ***cybernetic organism***. Cybernetics is the branch of information science that deals with the control of biological, mechanical, and/or electronic systems.

dark matter   Dark matter in the universe cannot be observed directly because it emits very little or no electromagnetic radiation. Scientists infer its existence through secondary phenomena such as gravitational effects and suggest that it may make up about 90 percent of the total mass of the universe. It is also called missing mass.

Deep Space Network (DSN)   NASA's global network of antennas serve as the radio-wave communications link to distant interplanetary spacecraft and probes, transmitting instructions to them and receiving data from them. Large radio antennas of the Deep Space Network's (DSN's) three Deep Space Communications Complexes (DSCCs) are located in Goldstone, California; near Madrid, Spain; and near Canberra, Australia. They provide almost continuous contact with a spacecraft in deep space as Earth rotates on its axis.

deep space probes   Spacecraft that are designed for exploring deep space, especially to the vicinity of the Moon and beyond, deep space probes include lunar probes, Mars probes, outer planet probes, solar probe, and so on.

Deimos   The tiny (about 7.5-miles [12-km] average diameter), irregularly shaped outer moon of Mars, Deimos was discovered in 1877 by the U.S. astronomer Asaph Hall (1829–1907).

Doppler shift   The apparent change in the observed frequency and wavelength of a source due to the relative motion of the source and an observer is a Doppler shift. If the source is approaching the observer, the observed frequency is higher and the observed wavelength is shorter. This change to shorter wavelengths is often called the blueshift. If the source is moving away from the observer, the observed frequency will be lower and the wavelength will be longer. This change to longer wavelengths is called the redshift. Named after the Austrian physicist Christian Johann Doppler (1803–53), who discovered this physical phenomenon in 1842 by observing sound waves.

downlink   A downlink is the telemetry signal that is received at a ground station from a spacecraft or space probe.

Drake equation   A probabilistic expression, proposed in 1961 by American astronomer Frank Donald Drake (1930–   ), the Drake equation is an interesting though highly speculative attempt to determine the number of advanced intelligent civilizations that might exist now in the Milky Way galaxy and be communicating (via radio waves) across interstellar distances. A basic assumption in Drake's formulation is the principle of mediocrity—namely that conditions in the solar system and even on Earth are nothing special particularly but, rather, represent common conditions found elsewhere in the galaxy.

dwarf planet   As defined by the International Astronomical Union (IAU) in August 2006, a dwarf planet is a celestial body that (a) is in orbit around the Sun, (b) has sufficient mass for its self-gravity to overcome rigid body forces so that it assumes a nearly round shape, (c) has not cleared the cosmic neighborhood around its orbit, and (d) is not a satellite of another (larger) body. Included in this definition are: Pluto, Ceres (the largest asteroid), and 2003 UB313 (a large, distant Kuiper-belt object nicknamed Xena and now officially called Eris).

dwarf star   Any star that is a main-sequence star, according to the Hertz-sprung-Russell (H-R) diagram, is a dwarf star. Most stars found in the galaxy, including the Sun, are of this type and are from 0.1 to about 100 solar masses in size. However, when astronomers use the term *dwarf star*, they are not referring to white dwarfs, brown dwarfs, or black dwarfs, which are celestial bodies that are not in the collection of main-sequence stars.

Dyson sphere   The British-American theoretical physicist Freeman John Dyson (1923–   ) suggested that an intelligent alien species might construct a huge, artificial biosphere that completely encircled their parent star as an upper limit of technological growth and interplanetary expansion within that solar system. This giant structure, or Dyson sphere, would be formed probably by a swarm of artificial habitats and miniplanets that is capable of intercepting essentially all the radiant energy from the parent star.

Earth-like planet   An Earth-like planet travels around another star (that is, an extrasolar planet), orbits in a continuously habitable zone (CHZ), and maintains environmental conditions resembling Earth. These conditions include a suitable atmosphere, a temperature range that permits the retention of large quantities of liquid water on the planet's surface, and a sufficient quantity of radiant energy that strikes the planet's surface from the parent star. Scientists in astrobiology hypothesize that under such conditions, the chemical evolution and development of carbon-based life (as known on Earth) could also occur there.

Earth's trapped radiation belts   Two major belts (or zones) of very energetic atomic particles (mainly electrons and protons) are trapped by Earth's magnetic field hundreds of miles (km) above the atmosphere. These are also called the Van Allen belts after the American physicist James Alfred Van Allen (1914–2006), who discovered them in 1958.

ecosphere   The ecosphere is a continuously habitable zone (CHZ) around a main-sequence star of a particular luminosity in which a planet could support environmental conditions that are favorable to the evolution and continued existence of life. For the chemical evolution of Earth-like, carbon-based, living organisms, global temperature and atmospheric pressure conditions must allow the retention of liquid water on the planet's surface. A viable ecosphere might lie about 0.7 and 1.3 astronomical units from a star like the Sun. However, if all the surface water has completely evaporated (the runaway greenhouse effect) or if all the liquid water on the planet's surface has completely frozen (the ice catastrophe), then an Earth-like planet within this ecosphere could not (in all likelihood) sustain life.

electromagnetic radiation (EMR)   Electromagnetic radiation is made up of oscillating electric and magnetic fields and is propagated with the speed of light. Includes (in order of decreasing frequency) gamma rays, X-rays, ultraviolet (UV) radiation, visible radiation, infrared (IR) radiation, radar, and radio waves.

escape velocity   (common symbol: V_e) The minimum velocity that an object must acquire to overcome the gravitational attraction of a celestial body, the escape velocity for an object that was launched from the surface of Earth is approximately 7 miles per second (11.2 km/s), while the escape velocity from the surface of Mars is about 3 miles per second (5.0 km/s).

Europa   The smooth, ice-covered moon of Jupiter, discovered in 1610 by the Italian astronomer Galileo Galilei (1564–1642). Astrobiologists currently think that Europa may have a life-bearing liquid-water ocean beneath its frozen surface. *See also* GALILEAN SATELLITES.

European Space Agency (ESA)   This international organization promotes the peaceful use of outer space and cooperation among the European member states in space research and applications.

exoatmospheric   Exoatmospheric means occurring outside Earth's atmosphere; events and actions that take place at altitudes above about 62 miles (100 km).

exobiology   The branch of science concerned with the study of the living universe, especially the possibility and characteristics of extraterrestrial life, is called exobiology. The term *astrobiology* is also used.

exotheology   Exotheology refers to the organized body of opinions concerning the impact that space exploration and the (possible) discovery of life beyond the boundaries of Earth would have on contemporary terrestrial religions. On Earth, theology involves the study of the nature of God and the relationship of human beings to God. The discovery of simple alien life-forms on other worlds within this solar system would undoubtedly rekindle serious interest in one of the oldest philosophical questions that has puzzled people throughout history: Are human beings the only intelligent species in the universe?

extragalactic   Occurring, located, or originating beyond the Milky Way galaxy is extragalactic.

extragalactic astronomy   The branch of astronomy—extragalactic astronomy—started about 1930 and deals with everything in the universe outside of the Milky Way galaxy.

extrasolar   Extrasolar refers to events occurring, located, or originating outside of the solar system.

extrasolar planet   An extrasolar planet belongs to a star other than the Sun. There are two general methods that scientists are using to detect extrasolar planets: direct—involving a search for telltale signs of a planet's infrared emissions—and indirect—involving precise observation of any perturbed motion of the parent star or any periodic variation in the intensity or spectral properties of its light.

extraterrestrial   Extraterrestrial events occur, are located, or originate beyond planet Earth and its atmosphere.

extraterrestrial catastrophe theory   The extraterrestrial catastrophe theory hypothesizes that a large asteroid or comet struck Earth some 65 million years ago, causing global environmental consequences that annihilated more than 90 percent of all animal species then living—including the dinosaurs.

extraterrestrial civilizations   *See* Kardashev civilizations.

extraterrestrial contamination   The contamination of one world by life-forms, especially microorganisms, from another world is extraterrestrial. Taking Earth's biosphere as the reference, planetary contamination is called forward contamination, when an alien world is contaminated by contact with terrestrial organisms, and back contamination, when alien organisms are released into Earth's biosphere.

extraterrestrial life   These life-forms may have evolved independently of and now exist beyond the terrestrial biosphere.

extravehicular activity (EVA)   Extravehicular activites are conducted by an astronaut or cosmonaut in outer space or on the surface of another planet (or moon), outside of the protective environment of his/her aerospace vehicle, spacecraft, or lander. Astronauts and cosmonauts must put on spacesuits (which contain portable life-support systems) to perform EVA tasks.

extremophile   A hardy (terrestrial) microorganism—an extremophile—can exist under extreme environmental conditions, such as in frigid polar regions or boiling hot springs. Astrobiologists speculate that similar (extraterrestrial) microorganisms might exist elsewhere in this solar system, perhaps within subsurface biological niches on Mars or in a suspected liquid-water ocean beneath the frozen surface of Europa.

farside   The side of the Moon that never faces Earth is called the farside.

Fermi paradox—"Where are they?"   The Italian-American physicist Enrico Fermi (1901–54) helped create the nuclear age. This brilliant Nobel laureate is also credited with the popular speculative inquiry now commonly called the Fermi paradox. His inquiry involves the migration of an advanced alien civilization through the galaxy on a wave of exploration. In 1943, Fermi suggested that if the universe is 15 billion years old and if just one intelligent star-faring civilization developed, then within about 100 million years, that civilization should have diffused throughout the entire Milky Way galaxy—making its presence known. So, where are they?

flare (solar)   A bright eruption from the Sun's corona, an intense flare represents a major ionizing radiation hazard to astronauts traveling beyond Earth's magnetosphere through interplanetary space or while exploring the surface of the Moon or of Mars.

flyby   In an interplanetary or deep space mission, a flyby spacecraft passes close to its target celestial body (e.g., a distant planet, a moon, an asteroid, or a comet) but does not impact the target or go into orbit around it.

free fall   An object falls unimpeded in a gravitational field. For example, all the astronauts and objects inside an Earth-orbiting spacecraft experience a continuous state of free fall and appear weightless as the force of inertia counterbalances the force of Earth's gravity.

frequency (usual symbol: f or ν)   Frequency is the rate of repetition of a recurring or regular event or the number of cycles of a wave per second. For electromagnetic radiation, the frequency (ν) is equal to the speed of light (c) divided by the wavelength (λ). *See also* HERTZ.

g   The symbol g is used for the acceleration due to gravity. At sea level on Earth, g is approximately 32.2 feet per second-squared (ft/s^2) (9.8 m/s^2)—that is, "one g." This term is used as a unit of stress for bodies that are experiencing acceleration. When a rocket accelerates during launch, everything inside it (including astronauts and cosmonauts) experiences a g-force that can be as high as several gs.

Gaia hypothesis   This hypothesis was first suggested in 1969 by the British biologist James Lovelock (1919–   )—with the assistance of the biologist Lynn Margulis (1938–   ): Earth's biosphere has an important modulating effect on the terrestrial atmosphere. Because of the chemical complexity observed in the lower atmosphere, Lovelock has suggested that life-forms within the terrestrial biosphere actually help to control the chemical composition of the Earth's atmosphere, thereby ensuring the continuation of conditions that are suitable for life.

galaxy   A galaxy is a very large accumulation of stars with from one million (10^6) to a million million (10^{12}) members. These island universes come in a variety of sizes and shapes, from dwarf galaxies (like the Magellanic Clouds) to majestic spiral galaxies (like the Andromeda galaxy). Astronomers classify them as elliptical, spiral (or barred spiral), or irregular.

Galilean satellites   The four largest and brightest moons of Jupiter, discovered in 1610 by the Italian astronomer Galileo Galilei (1564–1642), the Galilean satellites are Io, Europa, Ganymede, and Callisto.

Galileo Project   NASA's highly successful scientific mission to Jupiter was launched in October 1989. With electricity supplied by two radioisotope-thermoelectric generator (RTG) units, the *Galileo* spacecraft extensively studied the Jovian system from December 1995 until February 2003. On arrival, it also released an probe into the upper portions of Jupiter's atmosphere. On February 28, 2003, the NASA flight team terminated its operation of the *Galileo* spacecraft and commanded the robot craft to plunge into Jupiter's atmosphere. This mission-ending plunge took place in late September 2003.

gamma ray (symbol: γ)   Gamma rays are very short wavelength, high frequency packets (or quanta) of electromagnetic radiation. Gamma-ray photons are similar to X-rays except that they originate within the atomic nucleus and have energies between 10,000 electron volts (10 keV) and 10 million electron volts (10 MeV) or more.

Ganymede   *See* GALILEAN SATELLITES.

geocentric   Geocentric is relative to Earth as the center; it is measured from the center of Earth.

giant-impact model   The giant-impact model hypothesizes that the Moon originated when a Mars-sized object struck a young Earth with a glancing blow. The giant (oblique) impact released material that formed an accretion disk around Earth out of which the Moon formed.

giant planets   In this solar system, the large, gaseous outer planets are considered to be giant planets—Jupiter, Saturn, Uranus, and Neptune—as are any detected or suspected extrasolar planets that are as large or larger than Jupiter.

***Giotto* spacecraft**   Scientific robot spacecraft that was launched by the European Space Agency (ESA) in July 1985, the *Giotto* spacecraft successfully encountered the nucleus of comet Halley in mid-March 1986 at a (closest approach) distance of about 370 miles (600 km).

gravitation   Gravitation is the force of attraction between two masses. From Sir Isaac Newton's law of gravitation, this attractive force operates along a line joining the centers of mass, and its magnitude is inversely proportional to the square of the distance between the two masses. From Albert Einstein's general relativity theory, gravitation is viewed as a distortion of the space-time continuum.

gravity   Gravity is the attraction of a celestial body for any nearby mass, for example, the downward force imparted by Earth on a mass near Earth or on the planet's surface.

gravity assist   The change in a spacecraft's direction and speed that is achieved by a carefully calculated flyby through a planet's gravitational field, this change—or gravity assist—in spacecraft velocity occurs without the use of supplementary propulsive energy.

Hadley Rille   The long, ancient lava channel on the Moon that was the landing site for the *Apollo 15* mission during NASA's Apollo Project is called the Hadley Rille.

half-life (radioactive)   Radioactive half-life is the time required for one half of the atoms of a particular radioactive isotope population to disintegrate to another nuclear form. Measured half-lives vary from millionths of a second to billions of years.

halo orbit   A circular or elliptical orbit—a half orbit—in which a spacecraft remains in the vicinity of a Lagrangian libration point.

hard landing   A relatively high-velocity impact of a lander spacecraft on a solid planetary surface, this hard landing usually destroys all equipment, except perhaps a very rugged instrument package or payload container.

heliocentric   Heliocentric describes anything that has the Sun as a center.

heliopause   The boundary of the heliosphere, the heliopause is thought to occur about 100 astronomical units from the Sun and marks the edge of the Sun's influence and the beginning of interstellar space.

heliosphere   The region of outer space within the boundary of the heliopause in which the solar wind flows, the heliopause contains the Sun and the solar system.

hertz (symbol: Hz)   Hertz is the SI unit of frequency. One hertz is equal to one cycle per second. It was named in honor of the German physicist Heinrich Rudolf Hertz (1857–94), who produced and detected radio waves for the first time in 1888.

Hertzsprung-Russell (H-R) diagram   This useful graphic depiction of the different types of stars, arranged according to their spectral classification and luminosity, was named in honor of the Danish astronomer Ejnar Hertzsprung (1873–1967) and the American astronomer Henry Norris Russell (1877–1957), who developed the diagram independently of one another.

high Earth orbit (HEO)   A high Earth orbit circumnavigates the Earth at an altitude that is greater than 3,475 miles (5,600 km).

highlands   The oldest exposed areas on the surface of the Moon, the highlands are extensively cratered and chemically distinct from the maria.

Hohmann transfer orbit   The most efficient orbit transfer path between two coplanar circular orbits, the Hohmann transfer orbit consists of two impulsive high thrust burns (or firings) of a spacecraft's propulsion system. The technique was suggested in 1925 by the German engineer Walter Hohmann (1880–1945).

hot Jupiter   A Jupiter-sized planet in another solar system that orbits close enough to its parent star to have a high surface temperature is called a hot Jupiter.

"housekeeping" (spacecraft)   "Housekeeping" is the collection of routine tasks that must be performed to keep a spacecraft functioning properly during an orbital flight or interplanetary mission.

human-factors engineering   This branch of engineering is involved in the design, development, testing, and construction of devices, equipment, and artificial living environments to the anthropometric, physiological, and/or psychological requirements of the human beings who will use them. One aerospace example is the design of a functional microgravity toilet that is suitable for use by both male and female crewpersons.

***Huygens* probe**   The scientific *Huygens* probe was sponsored by the European Space Agency (ESA) and was named after the Dutch astronomer Christiaan Huygens (1629–95). The *Cassini* mother spacecraft delivered *Huygens* to Saturn, and the probe successfully plunged into the nitrogen-rich atmosphere of Titan (Saturn's largest moon) on January 14, 2005.

HZE particles   These are potentially the most damaging cosmic rays, with a high atomic number (Z) and high kinetic energy (KE).

Imbrium basin   The large (about 810 miles [1,300 km] across), ancient impact crater on the Moon is the Imbrium basin.

"infective theory of life"   The "infective theory of life" is the belief that some primitive form of life—perhaps selected, hardy bacteria or bioengineered microorganisms—was placed on an ancient Earth by members of a technically advanced extraterrestrial civilization.

infrared astronomy   This branch of astronomy deals with infrared (IR) radiation from relatively cool celestial objects, such as interstellar clouds of dust and gas (typically 100 K) and stars with surface temperatures below about 6,000 K.

International Space Station (**ISS**)   This major human space-flight project is headed by NASA. Russia, Canada, Europe, Japan, and Brazil are also contributing key elements to this large, modular space station in low Earth orbit that represents a permanent human outpost in outer space for microgravity research and advanced space technology demonstrations. On-orbit assembly began in December 1998.

international system of units   *See* SI UNITS.

interplanetary   Between the planets, within the solar system.

interplanetary dust (IPD)   Interplanetary dust consists of tiny particles of matter (generally less than 100 micrometers [μm] in diameter) found in outer space within the confines of this solar system.

interstellar   Between or among the stars.

interstellar communication and contact   Several methods of achieving contact with (*postulated*) intelligent extraterrestrial beings have been suggested. These include: (1) interstellar travel by means of starships, leading to physical contact between different civilizations; (2) indirect contact through the use of robot interstellar probes; (3) serendipitous contact, such as finding a derelict alien starship or probe drifting in the outer solar system; (4) interstellar communication involving the transmission and reception of electromagnetic signals; and (5) very "exotic" techniques involving information transfer through the modulation of gravitons, neutrinos, or possibly distortions in the space-time continuum.

interstellar medium (ISM)   The gas and tiny dust particles that are found between the stars in the Milky Way galaxy. Over 100 different types of molecules have been discovered in interstellar space, including many organic molecules.

interstellar probe   An interstellar probe is a conceptual, highly automated, robot spacecraft that has been launched by human beings in this solar system (or perhaps by intelligent alien beings in some other solar system) to explore nearby star systems.

intravehicular activity (IVA)   Astronaut or cosmonaut activities that are performed inside an orbiting spacecraft or aerospace vehicle are considered to be intravehicular. *Compare with* EXTRAVEHICULAR ACTIVITY.

Io   *See* Galilean satellites.

ionizing radiation   Any type of atomic or nuclear radiation that displaces electrons from atoms or molecules, thereby producing ions within the irradiated material, is said to be ionizing radiation. Examples include: alpha (α) radiation, beta (β) radiation, gamma (γ) radiation, protons, neutrons, and X-rays.

island universe   Term coined in the 18th century by the German philosopher Immanuel Kant (1724–1804), an island universe describes a distant collection of stars—now called a galaxy.

Jovian planet   A large (Jupiter-like) planet characterized by a great total mass, low average density, mostly liquid interior, and an abundance of the lighter elements (especially hydrogen and helium). In this solar system, the Jovian planets are Jupiter, Saturn, Uranus, and Neptune.

Kardashev civilizations   In 1964, the Russian astronomer Nikolai Semenovich Kardashev (1932–) examined the issue of information transmission by extraterrestrial civilizations and then postulated three types of technologically developed civilizations based on their energy use. A Type I civilization would be capable of harnessing the total energy capacity of its home planet; a Type II civilization harnesses the energy output of its parent star; and a Type III civilization would be capable of using and manipulating the energy output of the entire galaxy. *See also* DYSON SPHERE.

Kepler Mission   Scheduled for launch in 2008, NASA's *Kepler* spacecraft carries a unique space-based telescope that was specifically designed to search for earth-like planets around stars beyond the solar system.

Kuiper belt   The Kuiper belt is a region in the outer solar system beyond Neptune that extends out to perhaps 1,000 astronomical units and that contains millions of icy planetesimals. These icy objects range in size from tiny particles to Pluto-sized planetary bodies. The Dutch-American astronomer Gerard Peter Kuiper (1905–73) first suggested the existence of this disk-shaped reservoir of icy objects in 1951. *See also* OORT CLOUD.

laboratory hypothesis   A variation of the zoo hypothesis response to the Fermi paradox, this particular hypothesis postulates that the reason scientists cannot detect or interact with technically advanced extraterrestrial civilizations in the Milky Way galaxy is because they have set the solar system up as a "perfect" laboratory. *See also* FERMI PARADOX; ZOO HYPOTHESIS.

Lagrangian libration point   One of five points in outer space (called *L1, L2, L3, L4,* and *L5*) where a small object can experience a stable orbit in spite of the force of gravity exerted by two much-more-massive celestial bodies when they orbit about a common center of mass, the existence and location of these points were calculated by Joseph-Louis Lagrange (1736–1813) in 1772.

lander (spacecraft)   A lander is a spacecraft that is designed to reach the surface of a planet or moon safely and to survive long enough on the planetary body to collect useful scientific data that it sends back to Earth by telemetry.

life support system (LSS)   This system maintains life throughout the entire aerospace flight environment, including (as appropriate) travel in outer space, activities on the surface of another world (e.g., the lunar surface), and ascent and descent through Earth's atmosphere. The LSS must reliably satisfy a human crew's daily needs for clean air, potable water, food, and effective waste removal.

light-year (symbol: ly)   The distance that light (or other forms of electromagnetic radiation) travels in one year, one light-year equals a distance of approximately 5.87×10^{12} miles (9.46×10^{12} km) or 63,240 astronomical units (AU).

little green men (LGM)   A popular expression (originating in science-fiction literature) for extraterrestrial beings, presumably intelligent, is little green men.

low Earth orbit (LEO)   A circular orbit just above Earth's sensible atmosphere at an altitude of between 185 to 250 miles (300 to 400 km) is a low Earth orbit.

Lowell, Percival (1855–1916)   Late in the 19th century, the wealthy U.S. astronomer Percival Lowell established a private astronomical observatory (called the Lowell Observatory) near Flagstaff, Arizona—primarily to support his personal interest in Mars and his aggressive search for signs of an intelligent civilization there.

luminosity (symbol: L)   The rate at which a star or other luminous object emits energy, usually in the form of electromagnetic radiation, is called luminosity.

Luna   Luna is a series of robot spacecraft that was sent to explore the Moon in the 1960s and 1970s by the former Soviet Union.

lunar   Anything of or pertaining to Earth's natural satellite, the Moon, is termed lunar.

lunar base   A permanently inhabited complex on the surface of the Moon, a lunar base is the next logical step after brief human exploration expeditions such as NASA's Apollo Project.

lunar excursion module (LEM)   This lander spacecraft was used by NASA to deliver astronauts to surface of Moon during the Apollo Project.

lunar highlands   The light-colored, heavily cratered mountainous part of the Moon's surface is called the lunar highlands.

lunar orbiter   A spacecraft that is placed in orbit around the Moon, specifically, the lunar orbiter refers to the series of five *Lunar Orbiter* robot spacecraft that NASA used from 1966 to 1967 to photograph the Moon's surface precisely in support of the Apollo Project.

lunar probe   A planetary lunar probe explores and reports conditions on or about the Moon.

Lunar Prospector   This NASA orbiter spacecraft circled the Moon from 1998 to 1999, searching for mineral resources. Data collected by this robot spacecraft suggest the possible presence of water-ice deposits in the Moon's permanently shadowed polar regions.

lunar rover   A crewed or automated (robot) rover vehicle was used to explore the Moon's surface. NASA's lunar rover vehicle (LRV) served as a Moon car for Apollo Project astronauts during the *Apollo 15, 16,* and *17* expeditions. Russian *Lunokhod 1* and *2* robot rovers were operated on Moon from Earth between 1970 and 1973.

Lunokhod   A Russian eight-wheeled robot vehicle, Lunokhod was controlled by radio-wave signals from Earth and was used to perform lunar surface exploration during the *Luna 17* (1970) and *Luna 21* (1973) missions to the Moon.

magnetosphere   This is the region around a planet in which charged atomic particles are influenced (and often trapped) by the planet's own magnetic field rather than the magnetic field of the Sun as projected by the solar wind.

manned   An aerospace vehicle or system is occupied by one or more persons, male or female. The terms *crewed, human,* or *personed* are preferred to *manned* today in the aerospace literature. For example, a "manned mission to Mars" should be called a "human mission to Mars."

maria (singular: mare)   The Latin word for "seas," *maria* originally was used by the Italian astronomer Galileo Galilei (1564–1642) to describe the large, dark ancient lava flows on the lunar surface; he and other 17th-century astronomers thought these features were bodies of water on the Moon's surface. Following tradition, this term is still used by modern astronomers.

Mariner   A series of NASA planetary exploration robot spacecraft that performed important flyby and orbital missions to Mercury, Mars, and Venus in the 1960s and 1970s were dubbed Mariner.

Mars base   This is the surface base that will be needed to support human explorers during a Mars expedition later this century.

Mars expedition   The first crewed mission to visit Mars in this century will be known as the Mars expedition. Current concepts suggest a 600- to 1,000-day duration mission (starting from Earth orbit), a total crew size of as many as 15 astronauts, and about 30 days for surface excursion activities on Mars.

Mars Exploration Rover (MER) 2003 mission   In 2003, NASA launched identical twin Mars rovers that were designed to operate on the surface of the Red Planet. *Spirit* (MER-A) was launched from Cape Canaveral on June 10, 2003, and successfully landed on Mars on January 4, 2004. *Opportunity* (MER-B) was launched from Cape Canaveral on July 7, 2003, and successfully landed on Mars on January 25, 2004. Both soft landings used the airbag bounce-and-roll arrival that was demonstrated during the Mars

Pathfinder mission. *Spirit* landed in Gusev Crater, and *Opportunity* landed at Terra Meridiania. As of January 31, 2007, both rovers were still functioning. Despite a nonfunctioning right wheel, *Spirit* remains healthy and on the move in the Gusev Crater region. The fully functioning *Opportunity* is providing panoramic (surface-view) images of the Victoria Crater region of the Red Planet.

Mars Global Surveyor (MGS)   A NASA orbiter spacecraft, the *Mars Global Surveyor* was launched in November 1996 and performed detailed studies of the Martian surface and atmosphere since March 1999. The *Mars Global Surveyor* stopped communicating with scientists on Earth on November 2, 2006.

Mars Odyssey   Launched from Cape Canaveral by NASA in April 2001, the *Mars Odyssey 2001* is an orbiter spacecraft that was designed to conduct a detailed exploration of Mars with emphasis being given to the search for geological features that would indicate the presence of water—flowing on the surface in past or currently frozen in subsurface reservoirs. The spacecraft's primary science mission continued through August 2004 and, as of January 2007, *Odyssey* was functioning in an extended mission, which included service as a communications relay for the Mars Exploration Rovers (*Spirit* and *Opportunity*).

Mars Pathfinder   This innovative NASA mission successfully landed a Mars surface rover—a small robot called *Sojourner*—in the Ares Vallis region of the Red Planet in July 1997. For more than 80 days, human beings on Earth used teleoperation and telepresence to drive the six-wheeled minirover cautiously to interesting locations on the Martian surface.

Mars Reconnaissance Orbiter (MRO)   On August 12, 2005, NASA launched the *Mars Reconnaissance Orbiter* from Cape Canaveral Air Force Station, Florida. The spacecraft reached Mars on March 10, 2006. After a six-month long process of adjusting and trimming the shape of its orbit around the Red Planet, *MRO* started to perform its high-resolution mapping mission of the surface.

Martian   Of or relating to the planet Mars is termed *Martian*.

Martian meteorites   This refers to the collection of a dozen or so unusual meteorites that are considered to represent pieces of Mars that were blasted off the Red Planet by ancient impact collisions, wandered through space for millions of years, and eventually landed on Earth. In 1996, NASA scientists suggested that one particular specimen, called ALH84001, might contain fossilized evidence showing that primitive life may have existed on Mars more than 3.6 billion years ago.

metric system   *See* SI UNITS.

Milky Way galaxy   This is the humans' home galaxy—a large spiral galaxy that contains between 200 and 600 billion solar masses. The Sun lies some 30,000 light-years from the galactic center.

missing mass   *See* DARK MATTER.

moon   A small natural celestial body that orbits a larger one, a moon is a natural satellite.

Moon   Earth's only natural satellite and closest celestial neighbor, the Moon has an equatorial diameter of 2,159 miles (3,476 km), keeps the same side (nearside) toward Earth, and orbits at an average distance (center to center) of 238,758 miles (384,400 km).

mother spacecraft   A mother spacecraft is an exploration spacecraft that carries and deploys one or several atmospheric probes, lander spacecraft, and/or lander and rover spacecraft combinations when it arrives at a target planet. The mother spacecraft then relays data back to Earth and may also orbit the planet to perform its own scientific mission. NASA's *Galileo* spacecraft to Jupiter and *Cassini* spacecraft to Saturn are examples.

nadir   The direction from a spacecraft directly down toward the center of a planet, the nadir is the opposite of the ZENITH.

NASA   The National Aeronautics and Space Administration is the civilian space agency of the United States. Created in 1958 by an act of Congress, NASA's overall mission is to plan, direct, and conduct civilian (including scientific) aeronautical and space activities for peaceful purposes.

nearside   The side of the Moon that always faces Earth is the nearside.

nebula (plural: nebulas or nebulae)   A cloud of interstellar gas or dust, a nebula can be seen as either as a dark hole against a brighter background (called a dark nebula) or as a luminous patch of light (called a *bright nebula*).

New Horizons Pluto-Kuiper Belt Flyby   This reconnaissance-type exploration mission will help scientists to understand the icy worlds at the outer edge of the solar system. NASA launched this spacecraft from Cape Canaveral on January 19, 2006, and sent the robot probe on its long one-way mission to conduct a scientific encounter with the dwarf planet Pluto and its moon Charon (in 2015) and then to explore portions of the Kuiper belt that lies beyond. This spacecraft will help resolve some basic questions about the surface features and properties of these distant icy bodies as well as their geology, interior makeup, and atmospheres.

nova (plural: novas or novae)   From the Latin for "new," a nova is a highly evolved star that exhibits a sudden and exceptional brightness, usually temporary, and then returns to its former luminosity. A nova is now thought to be the outburst of a degenerate star in a binary star system.

nuclear radiation   Nuclear radiation is an ionizing radiation that consists of particles (such as alpha particles, beta particles, and neutrons), and very energetic electromagnetic radiation (that is, gamma rays). Atomic nuclei emit this type of radiation during a variety of energetic nuclear reaction processes, including radioactive decay, fission, and fusion.

nucleosynthesis   Nucleosynthesis is the production of heavier chemical elements from the fusion (joining together) of lighter chemical elements (such as, hydrogen and helium) in thermonuclear reactions in the interior of stars.

observable universe   The portions of the universe that can be detected and studied by the light they emit are called the observable universe.

Oort cloud   The large number (about 10^{12}) or cloud of comets that were postulated in 1950 by the Dutch astronomer Jan Hendrik Oort (1900–92), the Oort cloud orbits the Sun at an enormous distance—ranging from some 50,000 and 80,000 astronomical units.

orbiter (spacecraft)   An orbiter spacecraft is especially designed to travel through interplanetary space, achieve a stable orbit around the target planet (or other celestial body), and conduct a program of detailed scientific investigation.

orbiting quarantine facility (OQF)   An orbiting quarantine facility is a proposed Earth-orbiting, crew-tended laboratory in which soil and rock samples from Mars and other worlds in the solar system would first be tested for potentially harmful alien microorganisms—before these materials are allowed to enter Earth's biosphere.

outer space   Outer space is any region beyond Earth's atmospheric envelope—which usually is considered to begin at between 62 and 125 miles (100 and 200 km) altitude.

panspermia   The general hypothesis that microorganisms, spores, or bacteria that are attached to tiny particles of matter have diffused through space, eventually encountering a suitable planet and initiating the rise of life there, the word *panspermia* itself means "all-seeding."

parking orbit   The temporary (but stable) orbit of a spacecraft around a celestial body, a parking orbit is used for assembly and/or transfer of crew or equipment, as well as to wait for conditions favorable for departure from that orbit.

parsec (symbol: pc)   A parsec is an astronomical unit of distance that corresponds to a trigonometric parallax (π) of one second of arc. The term is a shortened form of *pa*rallax *sec*ond and 1 parsec represents a distance of 3.26 light-years (or 206,265 astronomical units).

perfect cosmological principle   The perfect cosmological principle is the postulation that at all times the universe appears the same to all observers. *See also* COSMOLOGY.

peri-   This prefix means near.

periastron   Periastron is the point of closest approach of two stars in a binary star system. *Compare with* APASTRON.

perilune   Perilune is the point in an elliptical orbit around the Moon that is nearest to the lunar surface. *Compare with* APOLUNE.

Phobos   The larger, innermost of the two small moons of Mars, Phobos was discovered in 1877 by the U.S. astronomer Asaph Hall (1829–1907). *See also* DEIMOS.

photometer   A photometer is an instrument that measures light intensity and the brightness of celestial objects, such as stars.

photosphere A photosphere is the intensely bright (white-light), visible surface of the Sun or other star.

Pioneer 10, 11 spacecraft NASA's twin robot spacecraft were the first to navigate the main asteroid belt, the first to visit Jupiter (1973 and 1974), the first to visit Saturn (*Pioneer 11* in 1979), and the first humanmade objects to leave the solar system (*Pioneer 10* in 1983). Each spacecraft is now on a different trajectory to the stars, carrying a special message (the Pioneer plaque) for any intelligent alien civilization that might find it millions of years from now.

planet A nonluminous celestial body that orbits around the Sun or some other star, the name *planet* comes from the ancient Greek *planetes* ("wanderers") since early astronomers identified the planets as the wandering points of light that are relative to the fixed stars. There are eight major planets, three dwarf planets, and numerous minor planets (or asteroids) in humans' solar system. In August 2006, the International Astronomical Union (IAU) clarified the difference between a planet and a dwarf planet. A planet is defined as a celestial body that (a) is in orbit around the Sun, (b) has sufficient mass for its self-gravity to overcome rigid body forces so as to assume a nearly round shape, and (c) has cleared the cosmic neighborhood around its orbit. Within this definition, there are eight major planets in the solar system: Mercury, Venus, Earth, Mars, Jupiter, Saturn, Uranus, and Neptune. Pluto is now regarded as a dwarf planet. *See also* DWARF PLANET.

planetary engineering Planetary engineering (also called terraforming) is the large-scale modification or manipulation of the environment of another planet to make it more suitable for human habitation. In the case of Mars, for example, human settlers would probably seek to make its atmosphere more dense and breathable by adding additional oxygen. Early Martian pioneers would most likely also attempt to alter the planet's harsh temperatures and modify them to fit a more terrestrial thermal pattern.

planetesimals Planetesimals are small rock and rock/ice celestial objects that are found in the solar system, ranging from 0.06 mile (0.1 km) to about 62 miles (100 km) in diameter.

planet fall The act of landing of a spacecraft or space vehicle on a planet or moon is called a planet fall.

polar orbit An orbit around a planet (or primary body) that passes over or near its poles, a polar orbit has an inclination of about 90°.

Population I stars These are hot luminous, young stars, including those like the Sun, that reside in the disk of a spiral galaxy and are higher in heavy element content (about 2 percent abundance) than Population II stars.

Population II stars Older stars, these are lower in heavy element content than Population I stars and reside in globular clusters as well as in the halo of a galaxy—that is, the distant spherical region that surrounds a galaxy.

primary body The primary body is a celestial body around which a satellite, a moon, or another object orbits, from which it is escaping, or toward which it is falling.

Project Cyclops Project Cyclops is a proposed, very large array of dish antennas for use in a detailed search of the radio-frequency spectrum (especially the 18- to 21-centimeter wavelength "water-hole" region of the spectrum) for interstellar signals from intelligent alien civilizations. The engineering details of this search for extraterrestrial intelligence (SETI) configuration were derived in a special summer-institute design study that was sponsored by NASA at Stanford University in 1971.

Project Daedalus This name was given to an extensive study of interstellar space exploration that was conducted from 1973 to 1978 by a team of scientists and engineers under the auspices of the British Interplanetary Society. The *Daedalus* robot spaceship, communications systems, and much of the payload were designed entirely within the parameters of 20th-century technology. The intended target of this proposed interstellar probe was Barnard's star, a red dwarf about 6 light-years away.

Project Ozma The pioneering attempt to detect interstellar radio-wave signals from an intelligent extraterrestrial civilization, the Project Ozma study was conducted by the U.S. astronomer Frank Donald Drake (1930–) in 1960 at the National Radio Astronomy Observatory in Green Bank, West Virginia. No strong evidence was found after 150 hours of listening for intelligent signals from the vicinity of two sunlike stars about 11 light-years away.

protogalaxy A protogalaxy is one that is in the early stages of evolution.

protoplanet A protoplanet is any of a star's planets as such planets emerge during the process of accretion in which planetesimals collide and coalesce into large objects.

protostar   A star in the making is a protostar. Specifically, this stage in a young star's evolution occurs after it has separated from a gas cloud but prior to it collapsing sufficiently (due to gravity) to support thermonuclear fusion reactions in its core.

Proxima Centauri   The closest star to the Sun—the third member of the Alpha Centauri triple-star system, Proxima Centauri is some 4.2 light-years away.

pulsar   A rapidly spinning neutron star that generates regular pulses of electromagnetic radiation, although originally discovered by radio-wave observations, pulsars have since been observed at optical, X-ray, and gamma-ray energies.

Quaoar   Large, icy world with a diameter of about 780 miles (1,250 km) located in the Kuiper belt about 1 million miles (1.6 million km) beyond Pluto, Quaoar was first observed in June 2004.

quasar   A mysterious, very distant object with a high redshift—that is, traveling away from Earth at great speed, a quasar will appear almost like a star but is far more distant than any individual star now observed. They might be the very luminous centers of active distant galaxies. When first identified in 1963, they were called *quasi*-stell*ar* radio sources—or quasars and are also called quasi-stellar objects (QSO).

radiation belt   This is the region(s) in a planet's magnetosphere where there is a high density of trapped atomic particles from the solar wind. *See also* EARTH'S TRAPPED RADIATION BELTS.

radio astronomy   This branch of astronomy collects and evaluates radio signals from extraterrestrial sources. Radio astronomy started in the 1930s when an American radio engineer, Karl Jansky (1905–50), detected the first extraterrestrial radio signals.

radio frequency (RF)   The radio-frequency portion of the electromagnetic spectrum is useful for telecommunications with a frequency range between 10,000 and 3×10^{11} hertz.

radio galaxy   A radio galaxy (often dumbbell shaped) produces very strong radio-wave signals. Cygnus A is an example of intense source about 650 million light-years away.

Ranger Project   The Ranger Project was the first NASA robot spacecraft that was sent to the Moon in the 1960s. These hard-impact planetary probes were designed to take a series of television images of the lunar surface before crash landing.

Red Planet   The Red Planet Mars is so named because of its distinctive reddish soil.

redshift   *See* Doppler shift.

regenerative life-support system (RLSS)   This is a controlled ecological life-support system in which biological and physiochemical subsystems produce plants for food and process solid, liquid, and gaseous wastes for reuse in the system.

regolith (lunar)   The unconsolidated mass of surface debris that overlies the Moon's bedrock is the regolith. This blanket of pulverized lunar dust and soil was created by millions of years of meteoric and cometary impacts.

relativity   The theory of space and time that was developed by the German-Swiss-U.S. physicist, Albert Einstein (1879–1955) early in the 20th century, relativity and the quantum theory serve as the two pillars of modern physics.

rendezvous   The close approach of two or more spacecraft in the same orbit, so that docking can take place, is a rendezvous. These objects meet at a preplanned location and time with essentially zero relative velocity.

robot spacecraft   A semiautomated or fully automated spacecraft that is capable of executing its primary exploration mission with minimal or no human supervision is a robot spacecraft.

rocket   A completely self-contained projectile or flying vehicle that is propelled by a reaction engine, a rocket carries all of its required propellant, so it can function in the vacuum of outer space, and it represents the key to space travel. There are chemical rockets, nuclear rockets, and electric propulsion rockets. Chemical rockets are further divided into solid-propellant rockets and liquid-propellant rockets.

rogue star   A rogue star is a wandering star that passes close to a solar system, disrupting the celestial bodies in the system and triggering cosmic catastrophes on life-bearing planets.

rover A crewed or robot space vehicle, a rover is used to explore a planetary surface.

satellite A secondary (smaller) celestial body, a satellite is in orbit around a larger primary body. For example, Earth is a natural satellite of the Sun, while the Moon is a natural satellite of Earth. A humanmade spacecraft placed in orbit around Earth is called an artificial satellite—or more commonly, just a satellite.

search for extraterrestrial intelligence (SETI) This is an attempt to answer the important philosophical question, Are we alone in the universe? The major objective of contemporary SETI programs (now being conducted by private foundations) is to detect coherent radio frequency (microwave) signals that are being generated by intelligent extraterrestrial civilizations—should they exist.

self-replicating system (SRS) An advanced robot system, the self-replicating system was first postulated by the Hungarian-born, German-American mathematician John von Neumann (1903–57). Space-age versions of the SRS would be very smart machines that are capable of gathering materials, performing self-maintenance, manufacturing desired products, and even making copies of themselves (self-replication).

sensor The sensor portion of a scientific instrument detects and/or measures some physical phenomenon.

shirt-sleeve environment A space-station module or spacecraft cabin in which the atmosphere is similar to that found on the surface of Earth; that is, a shirt-sleeve environment does not require a pressure suit.

SI units The international system of units (the metric system) that uses the meter (m), the kilogram (kg), and the second (s) as its basic units of length, mass, and time, respectively.

soft landing The act of landing on the surface of a planet without damaging any portion of a spacecraft or its payload, except possibly an expendable landing gear structure, is a soft landing. *Compare with* HARD LANDING.

Sol This is the Sun.

sol A Martian day (about 24 hours, 37 minutes, 23 seconds in duration), seven sols equal about 7.2 Earth days.

solar flare   A solar flare is a highly concentrated, explosive release of electromagnetic radiation and nuclear particles within the Sun's atmosphere near an active sunspot.

solar system   In general, a solar system is any star and its gravitationally bound collection of nonluminous objects, such as planets, asteroids, and comets; specifically, humans' home solar system consists of the Sun and all the objects that are bound to it by gravitation—including eight major planets, three dwarf planets (Pluto, Ceres, and Eris) with more than 60 known moons, more than 2,000 asteroids (minor planets), and a very large number of comets. Except for the comets, all the other celestial objects travel around the Sun in the same direction.

solar wind   The solar wind is the variable stream of plasma (that is, electrons, protons, alpha particles, and other atomic nuclei) that flows continuously outward from the Sun into interplanetary space.

space base   A large, permanently inhabited space facility—a space base—that is located in orbit around a celestial body or on its surface would serve as the center of future human operations in some particular region of the solar system.

space colony   An earlier term used to describe a large, permanent space habitat and industrial complex that is occupied by as many as 10,000 persons, currently the term *space settlement* is preferred to *space colony.*

spacecraft   A platform that can function, move, and operate in outer space or on a planetary surface, spacecraft can be human-occupied or uncrewed (robot) platforms. They can operate in orbit around Earth or while on an interplanetary trajectory to another celestial body. Some spacecraft travel through space and orbit another planet, while others descend to a planet's surface, making a hard landing (collision impact) or a (survivable) soft landing. Exploration spacecraft are often categorized as flyby, orbiter, atmospheric probe, lander, or rover spacecraft.

spacecraft clock   The time-keeping component within a spacecraft's command and data-handling system, a spacecraft clock meters the passing time during a mission and regulates nearly all activity within the spacecraft.

space debris   Space junk, abandoned or discarded humanmade objects in orbit around Earth, space debris includes operational debris (items discarded during spacecraft deployment), used or failed rockets, inactive or broken satellites, and fragments from collisions and space object breakup. When a spacecraft collides with an object or a discarded rocket spontaneously explodes, thousands of debris fragments become part of orbital debris population.

space launch vehicle (SLV)   A space launch vehicle is the expendable or reusable rocket-propelled vehicle that is used to lift a payload or spacecraft from the surface of Earth and place it in orbit around the planet or on an interplanetary trajectory.

spaceman   A person, male or female, who travels in outer space, the term *astronaut* is preferred to *spaceman*.

spaceport   A spaceport serves as both a doorway to outer space from the surface of a planet and a port of entry for aerospace vehicles returning from space to the planet's surface. NASA's Kennedy Space Center with its space shuttle launch site and landing complex is an example.

space radiation environment   One of the major concerns associated with the development of a permanent human presence in outer space is the ionizing radiation environment, both natural and humanmade. The natural portion of the space radiation environment consists primarily of Earth's trapped radiation belts (also called the Van Allen belts), solar particle events (SPEs), and galactic cosmic rays (GCRs).

space resources   The resources available in outer space that could be used to support an extended human presence and eventually become the physical basis for a thriving solar-system–level civilization, these resources include unlimited solar energy; minerals on the Moon, asteroids, Mars, and numerous outer planet moons; lunar (water) ice; and special environmental conditions such as access to high vacuum and physical isolation from terrestrial biosphere.

space settlement   A space settlement is a proposed, very large, humanmade habitat in outer space within which from 1,000 to 10,000 people would live, work, and play while supporting various research and commercial activities, such as the construction of satellite power systems.

spaceship   An interplanetary spacecraft that carries a human crew is a spaceship.

space shuttle The major space-flight component of NASA's Space Transportation System (STS), a space shuttle consists of a winged orbiter vehicle, three space-shuttle main engines (SSMEs), the giant external tank (ET)—which feeds liquid hydrogen and liquid oxygen to the shuttle's three main liquid propellant rocket engines—and the two solid rocket boosters (SRBs).

space sickness The space age form of motion sickness whose symptoms include nausea, vomiting, and general malaise, space sickness is a temporary condition that lasts no more than a day or so but affects 50 percent of the astronauts or cosmonauts when they encounter the microgravity environment (weightlessness) of an orbiting spacecraft after a launch. It is also called space-adaptation syndrome.

space station A space station is an Earth-orbiting facility designed to support long-term human habitation in outer space. *See also* INTERNATIONAL SPACE STATION.

spacesuit A flexible, outer garmentlike structure (including visored-helmet), a spacesuit protects an astronaut in the hostile environment of outer space, provides portable life-support functions, supports communications, and accommodates some level of movement and flexibility so that the astronaut can perform useful tasks during an extravehicular activity or while exploring the surface of another world.

Space Transportation System (STS) This is the official name for NASA's space shuttle.

space vehicle The general term, *space vehicle*, describes a crewed or robot vehicle that is capable of traveling through outer space. An aerospace vehicle can operate both in outer space and in Earth's atmosphere.

space walk The popular term for an extravehicular activity (EVA) is a space walk.

spectral classification In the spectral classification system, stars are given a designation that consists of a letter and a number, according to their spectral lines, which correspond roughly to surface temperature. Astronomers classify stars as O (hottest), B, A, F, G, K, and M (coolest). The numbers (0 through 9) represent subdivisions within each major class. The Sun is a G2 star—a little hotter than a G3 star and a little cooler than a G1 star. M stars are numerous but very dim, while O and B stars are very bright but rare.

spectroscopy   Spectroscopy is the study of spectral lines from different atoms and molecules. Astronomers use emission spectroscopy to infer the material composition of the objects that emitted the light and absorption spectroscopy to infer the composition of the intervening medium.

star   A self-luminous ball of very hot gas that liberates energy through thermonuclear fusion reactions within its core, a star is classified as either normal or abnormal. Normal stars, like the Sun, shine steadily—exhibiting one of a variety of distinctive colors such as red, orange, yellow, blue, and white (in order of increasing surface temperature). There are also several types of abnormal stars, including giant stars, white-, black-, and brown-dwarf stars, and variable stars. Stars experience an evolutionary life cycle from birth in an interstellar cloud of gas to death as a compact white dwarf, neutron star, or black hole.

star probe   A conceptual NASA robot scientific spacecraft, a star probe is capable of approaching within 620,000 miles (1,000,000 km) of the Sun's surface (photosphere) and providing the first in-situ measurements of its corona (outer atmosphere).

starship   A space vehicle—starship—is capable of traveling the great distances between star systems. Even the closest stars in the Milky Way galaxy are light-years apart. The term *starship* is generally used to describe an interstellar spaceship that is capable of carrying intelligent beings to other star systems; robot interstellar spaceships are often referred to as *interstellar probes.*

stationkeeping   The sequence of maneuvers that maintains a space vehicle or spacecraft in a predetermined orbit is known as stationkeeping.

sunlike star   A yellow, G spectral classification, main sequence star with a surface temperature between 5,000 and 6,000 K is called a sunlike star.

supernova   The catastrophic explosion, or supernova, of a massive star occurs at the end of its life cycle. As the star collapses and explodes, it experiences a variety of energetic nuclear reactions that lead to the creation of heavier elements, which are then scattered into space. Its brightness increases several million times in a matter of days and outshines all other objects in its galaxy.

Surveyor Project   The Surveyor Project was the NASA Moon exploration effort in which five lander spacecraft softly touched down on the lunar surface between 1966–68—the robot precursor to the Apollo Project human expeditions.

telecommunications　Telecommunications is the transmission of information over great distances using radio waves or other portions of the electromagnetic spectrum.

telemetry　The process of making measurements at one point and transmitting the information via radio waves over some distance to another location for evaluation and use is called telemetry. Telemetered data on a spacecraft's communications downlink often include scientific data as well as spacecraft state-of-health data.

teleoperation　The technique by which a human controller operates a versatile robot system that is at a distant, often hazardous, location is a teleoperation. High-resolution vision and tactile sensors on the robot, reliable telecommunications links, and computer-generated virtual reality displays enable the human worker to experience telepresence.

telepresence　The telepresence process is supported by an information-rich control station environment that enables a human controller to manipulate a distant robot through teleoperation and almost feel as if physically present in the robot's remote location.

telescope　An instrument that collects electromagnetic radiation from a distant object so as to form an image of the object or to permit the radiation signal to be analyzed, optical (astronomical) telescopes are divided into two general classes: refracting telescopes and reflecting telescopes. Earth-based astronomers also use large radio telescopes, while orbiting observatories use optical, infrared, ultraviolet, X-ray, and gamma-ray telescopes to study the universe.

terraforming　*See* PLANETARY ENGINEERING.

terrestrial　Terrestrial is of or relating to Earth.

terrestrial planets　In addition to Earth, the planets Mercury, Venus, and Mars—all of which are relatively small, high-density celestial bodies, are composed of metals and silicates with shallow or no atmospheres in comparison to the Jovian planets. These are the terrestrial planets.

Titan   The largest moon of Saturn, discovered in 1655 by the Dutch astronomer Christiaan Huygens (1629–95), Titan is the only moon in the solar system with a significant atmosphere.

transfer orbit   An elliptical interplanetary trajectory tangent to the orbits of both the departure planet and target planet (or moon) is a transfer orbit. *See also* HOHMANN TRANSFER ORBIT.

transit (planetary)   The passage of one celestial body in front of another (larger diameter) celestial body, such as Venus across the face of the Sun is a planetary transit.

Trans-Neptunian object (TNO)   Any of the numerous small, icy celestial bodies that lie in the outer fringes of the Solar System beyond Neptune, TNOs include plutinos and Kuiper belt objects.

unidentified flying object (UFO)   An unidentified flying object is seen (apparently) in the terrestrial skies by an observer who cannot determine its nature. The vast majority of such UFO sightings can, in fact, be explained by known phenomena. However, these phenomena may be beyond the knowledge or experience of the person who is making the observation. Common phenomena that have given rise to UFO reports include artificial Earth satellites, aircraft, high-altitude weather balloons, certain types of clouds, and the planet Venus. Since the late 1940s, there has been a popular (but unscientific and unfounded) association between UFO phenomena and alien spacecraft that are visiting Earth. The U.S. Air Force investigated such UFO phenomena from 1948 to 1969 under a variety of projects, including Project Blue Book.

uplink   The telemetry signal sent from a ground station to a spacecraft or planetary probe is an uplink.

Utopia Planitia   The smooth Martian plain on which NASA's *Viking 2* lander successfully touched down on September 3, 1976, is the Utopia Planitia.

Valles Marineris   An extensive canyon system on Mars near the planet's equator, Valles Marineris was discovered in 1971 by NASA's *Mariner 9* spacecraft.

Van Allen radiation belts   *See* EARTH'S TRAPPED RADIATION BELTS.

Very Large Array (VLA)   The Very Large Array (VLA) is a spatially extended radio telescope facility at Socorro, New Mexico. It consists of 27 antennas, each 82 feet (25 m) in diameter, that are configured in a giant "Y" arrangement on railroad tracks over a 12.4-mile (20-km) distance. The VLA is operated by the National Radio Astronomy Observatory and sponsored by the National Science Foundation.

Viking Project   In NASA's highly successful Mars exploration effort in the mid-1970s, the Viking Project 2 orbiter and 2 lander robot spacecraft conducted the first detailed study of the Martian environment and the first (albeit inconclusive) scientific search for life on the Red Planet.

Voyager   NASA's twin robot spacecraft Voyager explored the outer regions of the solar system, visiting all the Jovian planets. *Voyager 1* encountered Jupiter (1979) and Saturn (1980) before departing on an interstellar trajectory. *Voyager 2* performed the historic grand tour mission by visiting Jupiter (1979), Saturn (1981), Uranus (1986), and Neptune (1989). Both RTG-powered spacecraft are now involved in the Voyager Interstellar Mission (VIM) and each carries a special recording ("Sounds of Earth")—a digital message for any intelligent species that finds them drifting between the stars millennia from now.

water hole   The term *water hole* is used in the search for extraterrestrial intelligence (SETI) to describe a narrow portion of the electromagnetic spectrum that appears to be especially appropriate for interstellar communications between emerging and advanced civilizations. This band lies in the radio frequency (RF) part of the spectrum between 1,420 megahertz (MHz) frequency (21.1 cm wavelength) and 1,660 megahertz (MHz) frequency (18 cm wavelength).

white dwarf (star)   A compact star at the end of its life cycle, and once a star of one solar mass or less exhausts its nuclear fuel, a white dwarf star collapses under gravity into a very dense object about the size of Earth.

X-ray   An X-ray is a penetrating form of electromagnetic radiation of very short wavelength (approximately 0.01 to 10 nanometers) and high photon energy (approximately 100 electron volts to some 100 kiloelectron volts).

zenith   The zenith is the point on the celestial sphere that is vertically overhead. Compare with nadir, the point 180° from the zenith.

Zond   Zond is a family of robot spacecraft from the former Soviet Union that explored the Moon, Mars, Venus, and interplanetary space in the 1960s.

zoo hypothesis   One response to the Fermi paradox, the zoo hypothesis assumes that intelligent, very technically advanced species do exist in the Milky Way galaxy but that humans cannot detect or interact with them because they have set this solar system aside as a perfect zoo or wildlife preserve. *See also* FERMI PARADOX.

Further Reading

RECOMMENDED BOOKS

Angelo, Joseph A., Jr. *The Dictionary of Space Technology.* Rev. ed. New York: Facts On File, Inc., 2004.

———. *Encyclopedia of Space Exploration.* New York: Facts On File, Inc., 2000.

———, and Irving W. Ginsberg, eds. *Earth Observations and Global Change Decision Making, 1989: A National Partnership.* Malabar, Fla.: Krieger Publishing, 1990.

Brown, Robert A., ed. *Endeavour Views the Earth.* New York: Cambridge University Press, 1996.

Burrows, William E., and Walter Cronkite. *The Infinite Journey: Eyewitness Accounts of NASA and the Age of Space.* Discovery Books, 2000.

Chaisson, Eric, and Steve McMillian. *Astronomy Today.* 5th ed. Upper Saddle River, N.J.: Pearson Prentice Hall, 2005.

Cole, Michael D. *International Space Station. A Space Mission.* Springfield, N.J.: Enslow Publishers, 1999.

Collins, Michael. *Carrying the Fire.* New York: Cooper Square Publishers, 2001.

Consolmagno, Guy J., et al. *Turn Left at Orion: A Hundred Night Objects to See in a Small Telescope—And How to Find Them.* New York: Cambridge University Press, 2000.

Damon, Thomas D. *Introduction to Space: The Science of Spaceflight.* 3d ed. Malabar, Fla.: Krieger Publishing Co., 2000.

Dickinson, Terence. *The Universe and Beyond.* 3d ed. Willowdater, Ont.: Firefly Books Ltd., 1999.

Heppenheimer, Thomas A. *Countdown: A History of Space Flight.* New York: John Wiley and Sons, 1997.

Kluger, Jeffrey. *Journey beyond Selene: Remarkable Expeditions Past Our Moon and to the Ends of the Solar System.* New York: Simon & Schuster, 1999.

Kraemer, Robert S. *Beyond the Moon: A Golden Age of Planetary Exploration, 1971–1978.* Smithsonian History of Aviation and Spaceflight Series. Washington, D.C.: Smithsonian Institution Press, 2000.

Lewis, John S. *Rain of Iron and Ice: The Very Real Threat of Comet and Asteroid Bombardment.* Reading, Mass.: Addison-Wesley, 1996.

Logsdon, John M. *Together in Orbit: The Origins of International Participation in the Space Station.* NASA History Division, Monographs in Aerospace History 11, Washington, D.C.: Office of Policy and Plans, November 1998.

Matloff, Gregory L. *The Urban Astronomer: A Practical Guide for Observers in Cities and Suburbs.* New York: John Wiley and Sons, 1991.

Neal, Valerie, Cathleen S. Lewis, and Frank H. Winter. *Spaceflight: A Smithsonian Guide.* New York: Macmillan, 1995.

Pebbles, Curtis L. *The Corona Project: America's First Spy Satellites.* Annapolis, Md.: Naval Institute Press, 1997.

Seeds, Michael A. Horizons: *Exploring the Universe.* 6th ed. Pacific Grove, Calif.: Brooks/Cole Publishing, 1999.

Sutton, George Paul. *Rocket Propulsion Elements.* 7th ed. New York: John Wiley & Sons, 2000.

Todd, Deborah, and Joseph A. Angelo, Jr. *A to Z of Scientists in Space and Astronomy.* New York: Facts On File, Inc., 2005.

EXPLORING CYBERSPACE

In recent years, numerous Web sites dealing with astronomy, astrophysics, cosmology, space exploration, and the search for life beyond Earth have appeared on the Internet. Visits to such sites can provide information about the status of ongoing missions, such as NASA's *Cassini* spacecraft as it explores the Saturn system. This book can serve as an important companion, as you explore a new Web site and encounter a person, technology phrase, or physical concept unfamiliar to you and not fully discussed within the particular site. To help enrich the content of this book and to make your astronomy and/or space technology–related travels in cyberspace more enjoyable and productive, the following is a selected list of Web sites that are recommended for your viewing. From these sites you will be able to link to many other astronomy or space-related locations on the Internet. Please note that this is obviously just a partial list of the many astronomy and space-related Web sites now available. Every effort has been made at the time of publication to ensure the accuracy of the information provided. However, due to the dynamic nature of the Internet, URL changes do occur and any inconvenience you might experience is regretted.

Selected Organizational Home Pages

European Space Agency (ESA) is an international organization whose task is to provide for and promote, exclusively for peaceful purposes, cooperation among European states in space research and technology and their applications. URL: http://www.esrin.esa.it. Accessed on April 12, 2005.

National Aeronautics and Space Administration (NASA) is the civilian space agency of the United States government and was created in 1958 by an act of Congress. NASA's overall mission is to plan, direct, and conduct American civilian (including scientific) aeronautical and space activities for peaceful purposes. URL: http://www.nasa.gov. Accessed on April 12, 2005.

National Oceanic and Atmospheric Administration (NOAA) was established in 1970 as an agency within the U.S. Department of Commerce to ensure the safety of the general public from atmospheric phenomena and to provide the public with an understanding of Earth's environment and resources. URL: http://www.noaa.gov. Accessed on April 12, 2005.

National Reconnaissance Office (NRO) is the organization within the Department of Defense that designs, builds, and operates U.S. reconnaissance satellites. URL: http://www.nro.gov. Accessed on April 12, 2005.

United States Air Force (USAF) serves as the primary agent for the space defense needs of the United States. All military satellites are launched from Cape Canaveral Air Force Station, Florida or Vandenberg Air Force Base, California. URL: http://www.af.mil. Accessed on April 14, 2005.

United States Strategic Command (USSTRATCOM) is the strategic forces organization within the Department of Defense, which commands and controls U.S. nuclear forces and military space operations. URL: http://www.stratcom.mil. Accessed on April 14, 2005.

Selected NASA Centers

Ames Research Center (ARC) in Mountain View, California, is NASA's primary center for exobiology, information technology, and aeronautics. URL: http://www.arc.nasa.gov. Accessed on April 12, 2005.

Dryden Flight Research Center (DFRC) in Edwards, California, is NASA's center for atmospheric flight operations and aeronautical flight research. URL: http://www.dfrc.nasa.gov. Accessed on April 12, 2005.

Glenn Research Center (GRC) in Cleveland, Ohio, develops aerospace propulsion, power, and communications technology for NASA. URL: http://www.grc.nasa.gov. Accessed on April 12, 2005.

Goddard Space Flight Center (GSFC) in Greenbelt, Maryland, has a diverse range of responsibilities within NASA, including Earth system science, astrophysics, and operation of the *Hubble Space Telescope* and other Earth-orbiting spacecraft. URL: http://www.nasa.gov/goddard. Accessed on April 14, 2005.

Jet Propulsion Laboratory (JPL) in Pasadena, California, is a government-owned facility operated for NASA by Caltech. JPL manages and operates NASA's deep-space scientific missions, as well as the NASA's Deep Space Network, which communicates with solar system exploration spacecraft. URL: http://www.jpl.nasa.gov. Accessed on April 12, 2005.

Johnson Space Center (JSC) in Houston, Texas, is NASA's primary center for design, development, and testing of spacecraft and associated systems for human space flight, including astronaut selection and training. URL: http://www.jsc.nasa.gov. Accessed on April 12, 2005.

Kennedy Space Center (KSC) in Florida is the NASA center responsible for ground turnaround and support operations, prelaunch checkout, and launch of the space shuttle. This center is also responsible for NASA launch facilities at Vandenberg Air Force Base, California. URL: http://www.ksc.nasa.gov. Accessed on April 12, 2005.

Langley Research Center (LaRC) in Hampton, Virginia, is NASA's center for structures and materials, as well as hypersonic flight research and aircraft safety. URL: http://www.larc.nasa.gov. Accessed on April 15, 2005.

Marshall Space Flight Center (MSFC) in Huntsville, Alabama, serves as NASA's main research center for space propulsion, including contemporary rocket engine development as well as advanced space transportation system concepts. URL: http://www.msfc.nasa.gov. Accessed on April 12, 2005.

Stennis Space Center (SSC) in Mississippi is the main NASA center for large rocket engine testing, including space shuttle engines as well as future generations of space launch vehicles. URL: http://www.ssc.nasa.gov. Accessed on April 14, 2005.

Wallops Flight Facility (WFF) in Wallops Island, Virginia, manages NASA's suborbital sounding rocket program and scientific balloon flights to Earth's upper atmosphere. URL: http://www.wff.nasa.gov. Accessed on April 14, 2005.

White Sands Test Facility (WSTF) in White Sands, New Mexico, supports the space shuttle and space station programs by performing tests on and evaluating potentially hazardous materials, space flight components, and rocket propulsion systems. URL: http://www.wstf.nasa.gov. Accessed on April 12, 2005.

Selected Space Missions

Cassini Mission is an ongoing scientific exploration of the planet Saturn. URL: http://saturn.jpl.nasa.gov. Accessed on April 14, 2005.

Chandra X-ray Observatory (CXO) is a space-based astronomical observatory that is part of NASA's Great Observatories Program. *CXO* observes the universe in the X-ray portion of the electromagnetic spectrum. URL: http://www.chandra.harvard.edu. Accessed on April 14, 2005.

Exploration of Mars is the focus of this Web site, which features the results of numerous contemporary and previous flyby, orbiter, and lander robotic spacecraft. URL: http://mars.jpl.nasa.gov. Accessed on April 14, 2005.

National Space Science Data Center (NSSDC) provides a worldwide compilation of space missions and scientific spacecraft. URL: http://nssdc.gsfc.nasa.gov/planetary. Accessed on April 14, 2005.

Voyager (Deep Space/Interstellar) updates the status of NASA's *Voyager 1* and *2* spacecraft as they travel beyond the solar system. URL: http://voyager.jpl.nasa.gov. Accessed on April 14, 2005.

Other Interesting Astronomy and Space Sites

Arecibo Observatory in the tropical jungle of Puerto Rico is the world's largest radio/radar telescope. URL: http://www.naic.edu. Accessed on April 14, 2005.

Astrogeology (USGS) describes the USGS Astrogeology Research Program, which has a rich history of participation in space exploration efforts and planetary mapping. URL: http://planetarynames.wr.usgs.gov. Accessed on April 14, 2005.

Hubble Space Telescope **(HST)** is an orbiting NASA Great Observatory that is studying the universe primarily in the visible portions of the electromagnetic spectrum. URL: http://hubblesite.org. Accessed on April 14, 2005.

NASA's Deep Space Network (DSN) is a global network of antennas that provide telecommunications support to distant interplanetary spacecraft and probes. URL: http://deepspace.jpl.nasa.gov/dsn. Accessed on April 14, 2005.

NASA's Space Science News provides contemporary information about ongoing space science activities. URL: http://science.nasa.gov. Accessed on April 14, 2005.

National Air and Space Museum (NASM) of the Smithsonian Institution in Washington, D.C., maintains the largest collection of historic aircraft and spacecraft in the world. URL: http://www.nasm.si.edu. Accessed on April 14, 2005.

Planetary Photojournal is a NASA/JPL– sponsored Web site that provides an extensive collection of images of celestial objects within and beyond the solar system, historic and contemporary spacecraft used in space exploration, and advanced aerospace technologies URL: http://photojournal.jpl.nasa.gov. Accessed on April 14, 2005.

Planetary Society is the nonprofit organization founded in 1980 by Carl Sagan and other scientists that encourages all spacefaring nations to explore other worlds. URL: http://planetary.org. Accessed on April 14, 2005.

Search for Extraterrestrial Intelligence (SETI) Projects at UC Berkeley is a Web site that involves contemporary activities in the search for extraterrestrial intelligence (SETI), especially a radio SETI project that lets anyone with a computer and an Internet connection participate. URL: http://www.setiathome.ssl.berkeley.edu. Accessed on April 14, 2005.

Solar System Exploration is a NASA-sponsored and -maintained Web site that presents the last events, discoveries and missions involving the exploration of the solar system. URL: http://solarsystem.nasa.gov. Accessed on April 14, 2005.

Space Flight History is a gateway Web site sponsored and maintained by the NASA Johnson Space Center. It provides access to a wide variety of interesting data and historic reports dealing with (primarily U.S.) human space flight. URL: http://www11.jsc.nasa.gov/history. Accessed on April 14, 2005.

Space Flight Information (NASA) is a NASA-maintained and -sponsored gateway Web site that provides the latest information about human spaceflight activities, including the *International Space Station* and the space shuttle. URL: http://spaceflight.nasa.gov Accessed on April 14, 2005.

Index